Steve,

Very much appreciate your support. You offered ideas, comments and suggestions throughout the FDT programme.

We look forward to receiving your counsel for the next phase of the Openwealth transformation programme.

Best wishes

Gilbert.

11/2/14

Transforming Field and Service Operations

Gilbert Owusu • Paul O'Brien • John McCall •
Neil F. Doherty

Editors

Transforming Field and Service Operations

Methodologies for
Successful Technology-Driven
Business Transformation

Springer

Editors
Gilbert Owusu
Paul O'Brien
BT Technology,
Service & Operations
Martlesham Heath
United Kingdom

John McCall
The Robert Gordon University
Aberdeen
United Kingdom

Neil F. Doherty
Loughborough University
Loughborough
United Kingdom

ISBN 978-3-642-44969-7   ISBN 978-3-642-44970-3 (eBook)
DOI 10.1007/978-3-642-44970-3
Springer Heidelberg New York Dordrecht London

Library of Congress Control Number: 2014930011

Printed on acid-free paper

Springer is part of Springer Science+Business Media (www.springer.com)

Foreword

Field service operations lie at the heart of many service industries. For companies such as telecommunications providers or on-line retailers, the efficient and effective management of field resources is fundamental to realising excellent customer service and a competitive cost base. However, the successful delivery of the systems and processes that underpin field service operations is challenging. A study by the British Computer Society suggests that 'only around 16% of IT projects can be considered truly successful'.[1] This problem can be attributed to a number of factors, most notably failing to take a holistic approach to transforming operations. A successful approach combines technology change with process redesign, people engagement and organisational transformation.

This book provides an insight into how to successfully transform field service operations with automated technologies. It draws on years of experience from different industries and from different perspectives on realising change. This book captures a range of views from research and technology development, systems engineering and change management. The authors provide practical insights using case studies to highlight lessons learnt and areas for further research. We are sure the book will both inform and broaden the reader's understanding of field operations and how it can underpin business transformation.

CEO, BT France   Jerome Boillot
MD, Research & Innovation, BT   Tim Whitley

---

[1] 'The Challenge of Complex IT Projects', British Computer Society. London, 2004.

Introduction

The drive to realise operational efficiencies, improve customer service, develop new markets and accelerate new product introductions has substantially increased the complexity of field service operations. Historically, the imbalance between the supply and demand sides of the service delivery has been a challenge for service providers. Consumer demand for specific services may be uncertain and variable. At the same time, factors in supply planning are inflexible and slow to respond to changes in demand. In recent years, the strategic benefits of field service automation have become ever more critical to cost-effective service delivery. To maximise the efficiency and effectiveness of field service operations, organisations have embarked on a wide range of transformation programmes that have sought to introduce automation through the use of workforce management technologies. Despite the potential business value that can be delivered from such transformation programmes, too often, the automation technologies have not been fully utilised, and their expected benefits have not therefore been realised. Indeed, in many such instances, not only performance improvements have failed to materialise but serious problems have also arisen, in the form of increased structural costs, inadequate responses to customer requests and the failure to respond to new revenue opportunities.

Scholars of organisation change argue that the success of any transformation programme is a function of how well the technical, political, structural and social aspects of a specific project have been managed. Unfortunately, as these aspects are typically entangled in a highly complex and interdependent web, it is very difficult to fully understand and address each element effectively. Indeed, all too often, organisations will address one or two aspects to the exclusion of the others, and in so doing, a variety of undesirable consequences will ensue. Against this backdrop, the objective of this edited book is to provide insights into how organisations might successfully transform their field service operations with workforce management technologies. In addition to presenting a variety of case studies, which seek to demonstrate the lessons to be learnt from organisations' experiences of business transformation projects, this book also presents a range of practical tools, techniques and approaches that can be applied to facilitate service transformation.

Consequently, this book is aimed at those managers, technologists, change agents and scholars who are interested in field service operations. The book is organised into four parts.

Part I: The Case for Transforming Service and Field Operations

Together, the two chapters in the first part of this book aim to present the case as to why organisations should be actively considering transforming their service and field operations. Chapter 1, *IT Exploitation Through Business Transformation: Experiences and Implications*, sets the scene by presenting a fairly general exploration of how IT can be used to successfully facilitate business transformation, but only if the organisational change elements of the project are managed proactively. Its other key message is that the benefits from any such IT-facilitated business transformation project will only materialise, if the impacts of the technology are monitored and managed over its operational life: the realisation of business benefits is a journey, not a destination! By contrast, the following chapter, Chap. 2, *Transforming Field and Service Operations with Automation*, focuses far more explicitly on the effective transformation, in the context of field and service operations. In so doing, this chapter makes the case that organisation operating in a highly competitive sector, such as telecommunications, must transform their service operations if they are going to keep operating costs under control, whilst also maintaining high levels of customer satisfaction.

Part II: Methods, Models and Enabling Technologies for Transforming Service and Field Operations

There is a wide and diverse array of methods, models and enabling technologies that can be deployed with the aim of realising transformational improvements in service and field operations. In this part, several of these approaches are presented and evaluated. The techniques used include different forms of simulation at different operational levels and timescales, data-driven decision-making and analytics tools, along with computational modelling and fuzzy decision-making.

In Chap. 3, *Designing Effective Operations*, the authors present a cybernetic approach to identify key control loops and information flows in the management of complex operations. Organisational modelling and powerful simulation are used to gain insight by simplifying complex behaviours and making business assumptions and their impacts explicit. The use of these models by stakeholders can build a common ground for better balanced business decisions. The chapter provides a substantive example where a cybernetic simulation of the delivery of a telecom service is used to balance competing cost and customer service objectives.

Chapter 4, *System Dynamics Models of Field Operations*, focuses on long-term strategic planning. Here, there is a need to model the interaction between demand, resource levels and deployment and thus quantify trade-offs between cost and service performance. System dynamics modelling simulates the cumulative effect over time of operational decisions driven by the tension between competing performance objectives, targets and priorities. The types of scenarios that can be modelled are explained with a simple illustrative example. The chapter also provides a substantive real-world example of the application of system dynamics modelling to the field operations in a major telco.

In the age of big data, appropriate integration of data into operations management is increasingly automated, and so it is essential to adopt modern analytic approaches. These are the focus of Chaps. 5 and 6. Chapter 5, *Understanding the Risks of Forecasting*, explores the importance to forecasting of distinguishing between predictable variation, due to trends or seasonal effects, and inherence variation, or noise, in metrics relating to complex operations. The author provides several examples based on generated datasets to serve as a guide for forecasters, managers, modellers and process owners. Chapter 6, *Modern Analytics of Field Operations*, presents a variety of the tools and techniques designed by the authors. Businesses need to run their processes for field and service operations effectively and efficiently, providing good service at reasonable costs. Due to the changing nature of businesses including their environment and due to their intrinsic complexity, processes may require adaptation on a regular basis. Modern analytics can help improve processes and their execution by extracting the real process from workflow data (process mining), pointing to problems like bottlenecks and loops and by detecting emerging or changing patterns in demand and in the execution of processes (change pattern mining). The chapter explores these issues and gives examples of the successful application of the presented techniques.

Tactical supply planning (TSP) is intermediate between long-term strategic planning and the scheduling of day-to-day operations. The aim of TSP is to match the supply of resources to demand to balance cost and service levels. This is a complex task involving multiple skill sets and capacities and movement of resources between plans to meet shortages identified from demand forecasts. In Chap. 7, *Enhancing Field Service Operations via Fuzzy Automation of Tactical Supply Plan*, the authors review several different approaches to automate TSP process and present in detail a fuzzy logic-based approach. A key advantage of fuzzy logic is its ability to smooth the effects that changes in decision variable have on a cost function, avoiding the suboptimal effects of greedy approaches and so improving the overall quality of the plan.

Part III: Case Studies

A number of case studies help illustrate how new technologies can be applied to field and service operations to deliver business benefit. These cover three areas: decision support tools for helping to inform businesses on how to best realise transformation in field and service operations, core technologies for optimising resources used to deliver service and tools for improving performance in the field.

In Chap. 8, *The Role of Search for Field Force Knowledge Management*, the authors apply adaptive search techniques to improve the effectiveness of search in the field. This ensures the right information is available to the field engineer at the right time and shows the potential to significantly improve productivity. Equally important to the efficient management of a field force is the design of the activities they are performing in the field. Chapter 9, *Application of AI Methods to Practical GPON FTTH Network Design and Planning,* provides an insight into network planning optimisation and how AI technology can be employed to produce efficient FTTH (Fibre to the Home) network designs for a telecommunications network. An efficient design will reduce resources utilised in the field, such as manpower, network equipment, engineer travel time and energy consumed. Additionally, a semiautomated design tool reduces effort required in the network providers' planning offices. The support tools field engineers have access to also contribute to the efficient running of field and service operations and can help reduce task duration.

In Chap. 10, *The Role of Service Quality in Transforming Operations*, the authors critically assess the challenges in realising a service production management solution, before exploring how such systems can be used to effect important organisational changes. By contrast, in Chap. 11, *Field Force Management at Eircom,* a simulation tool for determining the most effective composition of the organisation's field service is presented. The simulation tool enables Eircom to explore what-if scenarios around factors such as skills, resource levels and geographical distribution which ensures that it optimally configures its field force. Such tools reduce the need to run expensive field trials in order to determine the efficacy of new approaches to field management. A more specialised modelling tool is then presented in Chap. 12, *Understanding Team Dynamics with Agent-Based Simulation*, which describes a dynamic business simulation environment. The capability enables the simulation of various working practices, organisational structures and work allocation scenarios in the light of event-driven perturbations to examine their impact on the execution of work plans based on historical or generated data. Chapter 13, *Effective Engagement of Field Service Teams*, outlines qualitative and quantitative requirements that will enable organisations to better understand how they can engage their employees and, in so doing, counter some of the negative effects, typically associated with significant transformation projects. The qualitative view, based on social research, reveals insights into peoples' actions, their behaviours, attitudes and values. The quantitative view provides the essential knowledge and information to facilitate the employee's job in hand as well as enable the workforce to reach full potential. In Chap. 14, *The Asset Replacement Problem: State of the Art*, the authors review the range of different modelling approaches available for determining asset replacement policy with fleet management being the major example used. Asset replacement is a complex decision process that must take into account the economic life of an asset, repair and replacement costs. The chapter reviews comprehensive cost minimisation models. In particular, it provides an analysis in detail of parallel replacement models and suggests a new model that addresses some of the issues not previously solved in this area. Finally, the limitations of the current models from a theoretical and applied perspective are discussed, and some remaining key challenges are identified for academics and practitioners working in this area.

Part IV: Challenges, Outcomes and Future Directions

Since the mid-1980s, concerted efforts have been geared towards exploiting IT in transforming service organisations. This motivation is underpinned by the belief that the benefits of IT have become ever more critical to cost-effective service delivery. These benefits can be classified along three business planning timeframes: strategic, operational and transactional. Typically, these benefits are measured as cost savings, cost avoidance or increase in revenue. Realising these benefits requires new models for communication across the service and changes in the behaviour of an organisation's resources and a seamless flow of information. In Chap. 15, *Enabling Smart Logistics for Service Operations*, the authors elucidate the concept of communication flexibility for improving information flows along the service chain. The provision of real-time data in the service chain highlights inefficiencies and opportunities for improving the accuracy of inventory measurement. Moreover, such insights can also be used to facilitate changes in operational procedures and cultures at an operational level. Successful change is about winning the minds and hearts of the people the change is intended for. Unfortunately, most IT transformation projects adopt a mechanistic approach which focuses predominately on the structural and symbolic aspects of the change with very little emphasis on the political and human resource parts. Enabling employees to reach their full potential using IT can be viewed as the desired outcome for any IT transformation project, i.e. realising and management of the benefits. In the final chapter, *Measuring and Managing the Benefits from IT Projects: A Review and Research Agenda*, the authors propose a new research agenda for addressing the key issues on benefits measurement, which it is hoped will facilitate the effective conduct of future transformation projects. More specifically, the research agenda includes the use of both quantitative and qualitative approaches to investigate application of benefits measurement at the strategic, operational and transactional decision-making in service organisations.

Martlesham Heath, UK   Gilbert Owusu
Aberdeen, UK   Paul O'Brien
Leicestershire, UK   John McCall
                                                                        Neil F. Doherty

Acknowledgements

The editors would like to thank the contributing authors. This book has benefited from their contributions and the diversity of their experience and viewpoints. We also want to express our sincere thanks to the following people for providing much valued feedback on the chapters that has improved the quality of the book—Mary Lumkin, David Lesaint, Alistair Duke, Raphael Dorne, Rob Claxton, Betul Bostanci and Chris Kirkbride. We are grateful to our colleagues who have supported the research, development and operationalising of service production management technologies at BT. Our heartfelt gratitude goes to all of them—Nader Azarmi, Chris Voudouris, Dave George, Nick Stork and the 'RM' team, Mark Ashpole, Chris Bilton, Ivan Boyd, Mathias Kern, Richard Starling, Bob Petty, Paul Cleaver and the FFA team, Yang Li, Michelle van der Peet, Richard Ward, Tony Frost, Richard Tunks, Biva Singh, Richard Burnham, Steve Buttery, Steve Hickey, Clive Ridsdale, Mike Elliott, Chris Hamilton, Nick Kennedy, Hari Srinivasan, Pani Kulamani, Ravi Guduru and the EWMP team, Kerry-Anne Lawlor, Gareth Douglas and the FDT team, Ian Elborn and the transformation team, Chris Cochrane and the CIO team. And finally, a special acknowledgement goes to Jerome Boillot and his team at BT France for their support and encouragement.

Contents

Part III Case Studies

Part IV Challenges, Outcomes and Future Directions

List of Contributors

Riaz Ahmad Etisalat Telecommunication Company, Abu Dhabi, UAE

Dyaa Albakour School of Computer Science, University of Glasgow, Glasgow, UK

Tanya Alcock Research and Innovation, BT Technology, Services and Operations, Martlesham Heath, UK

Amir H. Ansaripoor ESSEC Business School, Singapore, Singapore

Attracta Brennan GMIT, Mayo, Ireland

Nicola Buckhurst Research and Innovation, BT Technology, Services and Operations, Martlesham Heath, UK

Stephen A. Cassidy Research and Innovation, BT Technology, Services and Operations, Martlesham Heath, UK

Andrej Chu Etisalat-BT Innovation Centre (EBTIC), Khalifa University, Abu Dhabi, UAE

Crispin R. Coombs Centre for Information Management, Loughborough University, Leicestershire, UK

Neil F. Doherty Centre for Information Management, Loughborough University, Leicestershire, UK

Géry Ducatel Research and Innovation, BT Technology, Services and Operations, Martlesham Heath, UK

Leighton Evans University of Swansea, Swansea, UK

Hani Hagras School of Computer Science and Electronic Engineering, University of Essex, Essex, UK

Kjeld Jensen Research and Innovation, BT Technology, Services and Operations, Martlesham Heath, UK

Summer Kassem School of Computer Science and Electronic Engineering, University of Essex, Essex, UK

Udo Kruschwitz School of Computer Science and Electronic Engineering, University of Essex, Essex, UK

Anne Liret Research and Innovation, BT Technology, Services and Operations, Paris, France

Michael Lyons Research and Innovation, BT Technology, Services and Operations, Martlesham Heath, UK

Jonathan Malpass Research and Innovation, BT Technology, Services and Operations, Martlesham Heath, UK

Thierry Mamer School of Computing Science and Digital Media, Robert Gordon University, Aberdeen, UK

John McCall School of Computing Science and Digital Media, Robert Gordon University, Aberdeen, UK

Ahmed Mohamed School of Computer Science and Electronic Engineering, University of Essex, Essex, UK

Mohamed Naim Cardiff Business School, Cardiff, UK

Detlef Nauck Research and Innovation, BT Technology, Services and Operations, Martlesham Heath, UK

Irina Neaga School of Business and Economics, Loughborough University, Leicestershire, UK

Paul O'Brien Research and Innovation, BT Technology, Services and Operations, Martlesham Heath, UK

Fernando S. Oliveira ESSEC Business School, Paris, France

Anis Ouali Etisalat-BT Innovation Centre (EBTIC), Khalifa University, Abu Dhabi, UAE

Gilbert Owusu Research and Innovation, BT Technology, Services and Operations, Martlesham Heath, UK

Kin Fai (Danny) Poon Etisalat-BT Innovation Centre (EBTIC), Khalifa University, Abu Dhabi, UAE

Olivier Regnier-Coudert School of Computing Science and Digital Media, Robert Gordon University, Aberdeen, UK

Siddhartha Shakya Research and Innovation, BT Technology, Services and Operations, Martlesham Heath, UK

Martin Spott Research and Innovation, BT Technology, Services and Operations, Martlesham Heath, UK

Paul Taylor Research and Innovation, BT Technology, Services and Operations, Martlesham Heath, UK

Feargal Timon CIM Ireland, Galway, Ireland

Yingli Wang Cardiff Business School,, Cardiff, UK

David C. Wynn Engineering Design Centre, University of Cambridge, Cambridge, UK

Part I
The Case for Transforming Service and Field Operations

Chapter 1
IT Exploitation Through Business Transformation: Experiences and Implications

Neil F. Doherty and Crispin R. Coombs

Abstract In recent years it has been argued that as information technology has become a largely undifferentiated commodity, the scope for organisations to use it strategically, to gain and sustain a competitive advantage, has significantly diminished. In this chapter, we seek to assess the extent to which organisations will need to switch the focus of their attention from IT development to ongoing IT exploitation, if they wish to deliver real value from their commoditised IT solutions. In particular, we seek to explore the complex relationship between software projects, organisational change and benefits delivery, to try to understand why, if IT is now such a readily available commodity, it still results in such a wide variety of organisational impacts and outcomes. In so doing, this chapter will seek to articulate and critique the strategies and approaches that organisations will need to adopt if they are to be more successful in managing the business change and delivering the business benefits that stem from the adoption of commoditised IT.

1.1 Introduction

When Carr (2003) posed the question: *does IT matter*, he was raising legitimate concerns about whether information technology (IT) had become commoditised to the extent that it was generally perceived to be a utility that can be readily bought *off the shelf*, purely on the basis of cost and service performance. Indeed, there is now a great deal of evidence to suggest that many organisations are very keen to adopt standard systems such as accounting packages or ERP systems, on the basis of minimising costs and risks, even if it means sacrificing any opportunity for differentiation (Ravichandran and Liu 2011; Gilbert et al. 2012).

N.F. Doherty (✉) • C.R. Coombs
Centre for Information Management, Loughborough University, Leicestershire, UK
e-mail: n.f.doherty@lboro.ac.uk

G. Owusu et al. (eds.), *Transforming Field and Service Operations*,
DOI 10.1007/978-3-642-44970-3_1, © Springer-Verlag Berlin Heidelberg 2013

If IT is now perceived to be a largely undifferentiated commodity, then its ubiquity makes it, in effect, an equaliser—the same technology is available for purchase to everyone (Gilbert et al. 2012). The corollary of this conclusion is that one might expect any such commoditised technology to deliver similar organisational impacts and economic returns, irrespective of the organisational context in which it has been implemented and ultimately operated. However, there is a growing consensus within the literature that many of the human and organisational impacts of IT are certainly not deterministic and cannot therefore be easily predicted prior to a system's implementation (e.g. DeSanctis and Poole 1994; Leonardi 2007). Moreover, as many organisations have learnt to their cost, the economic returns and performance improvements to be realised from IT are a very long way from being uniform and predictable. It is widely acknowledged that a considerable amount of time, money, effort and opportunity has been wasted upon IT investments that have either been abandoned or ultimately failed to deliver any appreciable benefit (Fortune and Peters 2005). Indeed, it has been suggested that *only around 16 % of IT projects can be considered truly successful* (BCS 2004).

In this chapter, we aim to explore the extent to which organisations appear to be viewing IT as an undifferentiated commodity, the implications that this might have for the management of IT-induced organisational change (Eason 1988) and ultimately the realisation of benefits from IT investment projects. The remainder of this chapter is organised into four parts. First, we provide a brief but critical review of the growing literature that provides support for the view that IT is now a highly standardised commodity. We then review the literature to understand the types of organisational change that are engendered through the adoption of IT and the extent to which these impacts are predictable. In the fourth section, we focus upon the value that is leveraged from IT and the circumstances under which benefits might, or might not, be forthcoming. Finally, we seek to explain why IT delivers such unequal returns, and in so doing we argue that whilst IT might no longer matter, how IT is exploited over its operational life does matter.

1.2 The Commoditisation of Information Technology

Organisations seeking to achieve competitive advantage must typically develop strategies based upon differentiation, which give their activities a distinctiveness, which will ultimately be valued by customers (Enders et al. 2009). As IT is now such a readily accessible, affordable and homogenous commodity, it has been argued (Thatcher and Pingry 2007) that its potential to deliver any sustainable competitive advantage has become severely restricted, as technology, alone, can no longer make an organisation's activities distinctive. Although the organisational roles and impact of IT have changed dramatically in the last few decades, in many ways, IT is not dissimilar to other disruptive technologies that have previously transformed the industrial world (Carr 2003). It is widely acknowledged that IT may have provided a differentiated advantage to some companies early on, but over time IT has grown cheaper and more standardised so that it is easily accessible to everyone. The claim that 'IT no longer matters' resonates with the earlier *strategic necessity hypothesis* (Powell and Dent-Micallef 1997), which asserts that it is unlikely that any individual application of IT will be able to deliver a sustainable competitive advantage, because it is relatively easy for firms to understand, and then copy their competitors' systems, and that failure to do so will leave them competitively disadvantaged (Melville et al. 2004).

Against this backdrop, more and more organisations have tended to base their IT investment decisions on the dual criteria of cost minimisation and risk aversion (Gilbert et al 2012). For example, many organisations are seeking to attain far cheaper, faster and safer solutions by implementing readily available commercial off-the-shelf (COTS) solutions (Berg 2008). The rapid growth of outsourcing and shared service arrangements, in which common business systems and services are provided more cheaply through third-party providers (Chan and Gurbaxani 2012), also provides compelling evidence that organisations are going for cheaper and less risky solutions. More specifically, Wyld (2009) argues that cloud computing should ultimately allow organisations to treat business computing as a utility that is provided on an *on-demand* basis that is akin to the way in which businesses meet their need for electricity or telephony services. Many scholars (e.g. Ravichandran et al. 2009; Currie 2011) have demonstrated using *institutional theory* that growing numbers of organisations seek to reduce both costs and uncertainty by simply investing in the same types of technology as their competitors. As technology costs tend to decline with time, early investors in emerging technologies often pay higher prices for the technology. Consequently, firms that resist the temptation to aggressively invest in emerging technologies are likely to avoid significant risks and costs (Ravichandran and Liu 2011). If organisations are generally adopting a cost minimisation and risk aversion strategy, when it comes to their IT investment strategy, this begs the question of the extent to which this strategy has made the organisational impacts to be delivered by IT any more consistent and predictable.

1.3 The Organisational Impacts of IT

Information technology is now a ubiquitous and increasingly critical part of the fabric of the modern organisation, supporting its day-to-day operations and all aspects of the decision-making process, as well as, many would still argue, its strategic positioning (Nevo and Wade 2010). It is therefore not perhaps surprising that the implementation of a new technology or information system is often perceived to be one of the most effective ways of delivering significant changes to the design of an organisation (Markus 2004). At a macro level, such changes might be targeted on an organisation's structural arrangements, its culture or its key business processes. At a micro level, technology can be used to effect radical changes to working practices, user behaviours and job demarcations. For example, the implementation of a highly integrated and enterprise-wide system, such as ERP, can be explicitly used to modify:

- The design of clerical jobs and working practices (Marler and Lian 2012)
- The empowerment of workers (Esteves 2009; Murphy et al. 2012)
- The redesign of core business processes (Koch 2001; Fearon et al. 2013)
- The design of organisational structures and formal lines of reporting (Doherty et al. 2010)
- Managerial decision-making processes (Fearon et al. 2013)
- The distribution of power within organisations (de Vries and Boonstra 2012)
- The predominant value systems and cultural norms, within organisations (Waring and Skoumpopoulou 2012)

Against this backdrop, many commentators argue that the majority of benefits associated with an IT project stem indirectly from the resultant organisational change, rather than directly from the new functions and features offered by a newly installed piece of software (Ward and Elvin 1999). Consequently, project managers must explicitly seek to transform organisational structures and behaviours to ensure that the functionality of a newly implemented information system can be effectively leveraged through the design of the host organisation (Hughes and Scott Morton 2006; Peppard et al. 2007). Such technologically mediated organisational change programmes must be explicitly tailored to reflect the specific characteristics of the technology and the needs of the host organisation. As Clegg and Shepherd (2007) note, it is important to change the mindset of stakeholders, to reflect that they are primarily engaged in business transformation projects, rather than IT development projects. For example, benefits may only be leveraged from a new data warehouse if the host organisation actively seeks to modify its culture so that its staff feel free to act in a more flexible, customer-focussed and empowered manner (Markus 2004). In a similar vein, Peppard et al. (2007; p. 6) describe how benefits could only be leveraged from a new CRM system, implemented by a European paper manufacturer, by redefining the job descriptions and working practices, of sales representatives, so that they were *allocated more sales time to contact potential high-value customer leads.*

The literature is reasonably consistent in promoting the view that IT projects almost always induce significant organisational changes and that such change should be actively managed, to ensure that it either facilitates the leveraging of benefits from the technical artefact (Ashurst et al. 2008) or does not lead to dysfunctional user behaviours (Martinsons and Chong 1999). Unfortunately, although some of the impacts of IT may be relatively deterministic and predictable and are therefore amenable to proactive management, there is a growing body of work which argues that as the technical artefact is primarily a social construct (Grint and Woolgar 1997; Doherty et al. 2006), then many of its consequences and impacts will be difficult, if not impossible, to anticipate, in advance of implementation (Robey and Boudreau 1999). Whilst some of these unanticipated consequences, or incidental side effects, may be of a positive nature, negative impacts are also quite common, as IT-induced organisational change often results in user resistance and, in extreme cases, possibly even system rejection (Martinsons and Chong 1999). Consequently, although IT development projects should always be accompanied by proactive organisational change initiatives (Eason 2001), it is highly likely that an ongoing process of organisational adaptation will be also required, once the system is fully operational, to ensure that the functionalities of the IT artefact are being exploited to their full potential.

The view that the implementation of a new piece of business software engenders a wide variety of organisational impacts is still consistently held (Orlikowski 2010; Marler and Lian 2012; Doherty et al. 2012), but the extent to which these impacts are now more predictable, in the era of commoditised IT, needs to be further explored. In principle, it might seem reasonable to hypothesise that the organisational impacts of commoditised IT will now be far easier to predict, as there is now a considerable body of experience concerning the adoption of these technologies. Unfortunately, there is a great deal of evidence to suggest that the implementation of a highly standardised technology, such as ERP, still results in a very large number of unexpected organisational consequences and outcomes (Boudreau and Robey 2005; Staehr 2010; Murphy et al. 2012; Waring and Skoumpopoulou 2012). For example, Hanseth et al. (2006) relate the tale of how the implementation of a hospital-based ERP system, which was designed to create a single source of patient data and reduce the volume of paper-based records, resulted in completely the opposite outcome: a more fragmented patient record and increased volumes of paper records. Consequently, if the organisational impacts of IT are both critical to the downstream realisation of business benefits and still very difficult to forecast, prior to a system's implementation, then it is likely that the returns from an IT project will also be very difficult to predict.

1.4   Do IT Investment Projects Deliver Value?

As demonstrated earlier, there is already a significant body of evidence to suggest that more and more organisations are primarily basing their IT investment decisions on the twin criteria of cost minimisation and risk aversion (Carr 2003; Ravichandran and Liu 2011; Gilbert et al. 2012). However, there is no clear evidence to suggest that such a strategy will automatically deliver success. Although the current incidence of IT failures may have marginally improved from the period in which Clegg et al. (1997) reported that *up to 90 % of all IT projects fail to meet their goals*, the outcomes of IT projects are still highly variable. For example, in their study of more recent IT project outcomes, over the period 1994–2002, Shpilberg et al. (2007) reported that 74 % still failed to deliver their expected value. Moreover, an even more recent survey of IT executives found that 24 % of IT projects were still viewed as outright *failures*, whilst a further 44 % of projects were considered to be *challenged*, as they were finished late, over budget or with fewer than the required features and functions (Levinson 2009). Consequently, even if the majority of organisations are now opting for commoditised solutions, to reduce cost and avoid risk, there is little evidence to suggest that the outcomes will be wholly successful.

If IT executives do view IT as a ubiquitous and largely undifferentiated commodity, then it becomes very tempting for them to assume that it will automatically deliver value (Ashurst et al. 2008). In essence, they might be very prepared to presume that the responsibility for the success of their packaged or outsourced solution resides solely with the software provider or service provider, rather than themselves. Unfortunately, even highly packaged and commoditised solution can still prove to be troublesome. For example, Barker and Frolick (2003) describe how a major soft drink bottler successfully implemented an ERP system, which was intended to provide the benefit of integrated communication, but once live was considered to be a barrier to organisational effectiveness. In a similar vein, Peppard et al. (2007) report the case of a newly implemented CRM package that was delivered to time, budget and specification but provided no obvious benefits to the organisation. These studies show that if investments in commoditised IT are to be considered truly successful, then they have to achieve far more than simply being implemented to specification, on time and within budget (Dorgan and Dowdy 2004; Sauer and Davis 2010). Consequently, there may be a very substantial expectation's gap between the returns that an executive might expect from a newly implemented piece of commoditised software and the benefits that are ultimately realised.

1.5 Moving Towards an IT Exploitation Agenda

Whilst the returns from IT investments may often disappoint, there can be little argument that appropriate software has the potential to play some positive role in transforming the performance of business operations and enhancing the delivery of services. Consequently, from an organizational perspective, IT certainly can matter. However, when organizations are making specific decisions about which software to invest in, they might do well to heed the advice of Gilbert et al. (2012; p. 184), who concluded that '*the lesson from this study for practitioners, at least those at information technology-using industries, is to manage information technology to keep costs and risks under control and look elsewhere for innovation*'. Although organisations might be well advised to base their IT investment decisions on the basis of cost minimisation and risk reduction, we would argue that they still need to explicitly focus on strategies for leveraging value from such investments. Moreover, we would encourage them to still seek to use IT as a platform for innovation but not necessarily at the point of implementation.

It has been argued that the realisation of benefits from IT is *a journey not a destination* (Doherty 2014). In traditional systems development projects, the implementation of the software artefact tends to be the point at which most of the project activity, as well as any senior management interest, tends to wane (Peppard and Ward 2005). Unfortunately, from a benefits realisation perspective, this situation is seriously deficient, as benefits need to be actively managed over the system's operational life (Leonardi 2007). This longer-term exploitation strategy is often advantageous, as it encourages stakeholders to innovate and improvise with their local working environments (Orlikowski 1996), and to tailor their systems and processes, to reflect changing organisational circumstances and requirements. As Jasperson et al. (2005) note, organisations may be able to achieve considerable economic benefits (via relatively low incremental investment) by enabling users to enrich their use of already-installed information systems.

Unfortunately, it is not clear how easy it will be for organisations to leverage value from their IT investments, once operational, as relatively little attention has been devoted to examining how existing IT installations can be exploited by firms, to provide ongoing innovation opportunities. Much of the extant literature concerning the post-implementation use of IT has very narrowly focused upon the initial uptake and adoption of IT, rather than any long-term user behaviours (Ahuja and Thatcher 2005). Consequently, there is now a pressing need for wider research that goes beyond examining user acceptance behaviours of systems in the immediate post-adoption period and addresses the long-term exploitation of IT investments (Jasperson et al. 2005). To summarise, not only is the implementation of a new piece of software typically the signal for many IT professionals to move swiftly on to new challenges, it would also appear to be the point at which the interest of the majority of information system researchers quickly evaporates (Doherty 2014).

But what can be done to address this sorry state of affairs? The time would seem ripe for members of the practitioner and research communities to shift the focal point for the bulk of their work from pre-implementation activities to the ongoing refinement and exploitation of software once implemented. A research agenda to reflect this shift in emphasis would need to reflect the synergistic relationship between ongoing IT exploitation and business transformation, and it might productively focus on specific issues such as:

- **Proactive Benefits Management**: It is not enough to simply define all project activities in terms of the benefits to be delivered; their realisation has to be actively planned and managed throughout the system's operational life. Having identified the benefits to be delivered from a particular software project, managers will need to initiate an ongoing benefits realisation programme to ensure that all projected benefits are proactively managed, throughout the life of the project (Ward and Elvin 1999).
- **Benefits Exploitation**: In traditional systems development projects, the implementation of the software artefact tends to be the point at which most of the project activity, as well as any senior management interest, tends to wane (Peppard and Ward 2005). Unfortunately, from a benefits realisation perspective, this situation is seriously deficient, as benefits management needs to continue beyond the completion of the project and to be actively managed

throughout the system's operational life (Leonardi 2007). Consequently, the key focal point for reflecting upon benefits must shift from the planning and development phases to the system's operational phase.

- **Ongoing User Engagement**: Many systems fail to deliver their full potential either because they are suffering from problems, misinterpretations and bottlenecks, which do not become evident until the system is operational, or because important opportunities are being overlooked. In the future, organisations will need to actively engage users and other stakeholders in an ongoing dialogue, to critically appraise the performance of their systems and any associated business processes.

- **Encourage Innovation**: If members of a user community are encouraged to adopt, or simply fall into, a mindset that views commoditised IT as behaving in unidimensional manner, with little opportunity for alternative modes of operation, then it is very unlikely that they will use a technology to its full potential. Consequently, in the face of changing organisational circumstances and requirements, stakeholders should be encouraged to innovate and improvise with their local working environments (Orlikowski 1996) and to tailor their systems and processes to leverage performance improvements.

- **Business Process Redesign**: Having asked users to actively reflect on the performance of their systems and to think innovatively about how it can be improved, organisations will need to be far more prepared to make ongoing changes to the design of their business processes and the specifications of user behaviours.

- **Software Customisation**: As with ongoing business process redesign, organisations must also be prepared to customise their software, as and when necessary, to ensure that it consistently meets changing organisational requirements and opportunities.

- **User Training and Education**: The training and education of users should no longer be thought of as wholly pre-implementation activities. It will be essential for the long-term effectiveness of any organisational application of IT that users are encouraged to engage in ongoing training and education, to ensure that systems continue to be used to their full potential.

- **IT Capabilities**: Organisations will only be able to fully exploit their business technologies, if they make a substantive investment in developing the rich variety of in-house capabilities that are necessary to proactively manage the realisation of benefits from their complete portfolio of business systems throughout their operational lives (Leonardi 2007; Ashurst et al. 2008).

Whilst all these issues would be productive areas for researchers and practitioners, it is important to remember that these topics are not independent and the relationships between them are perhaps more important than individual issues. For example, what specific capabilities might organisations need to effectively customise their software to reflect job redesigns in order to deliver benefits?

1.6   Conclusions

IT professionals, academics and users are often tempted to refer to their software systems and applications as tools. Indeed, these are increasingly important tools, as they have the potential to leverage significant improvements to any organization's business processes, service operations and, in so doing, their operational performance. However, when other types of tool are put in the hands of an unfamiliar user, be it a chisel, a lathe or a scalpel, there is an automatic assumption that it will take months, if not years, of training, experimentation and practice, before he or she can use it to good effect. By contrast, when IT tools are deployed, there is often a wholly unrealistic expectation that they will immediately start to deliver organisational benefits and will continue to do so with little or no ongoing intervention or proactive support. In this chapter we argue that as IT becomes more commoditised, organisations should make a significant shift in their IT activities from the design, development and implementation of IT solutions to the exploitation of information systems through ongoing business transformation. Although we have attempted to identify some of potential mechanisms through which operational IT might best be exploited, the primary purpose of this chapter is to encourage IT researchers and executives to shift their focus from IT development to IT exploitation. Finally, It should be noted that many of the modelling and simulation tools, described elsewhere in this book, can be used to help understand the effectiveness of IT-enabled processes, and in so doing, help to focus an organization's IT exploitation activities.

References

Ahuja MK, Thatcher J (2005) Moving beyond intentions and toward the theory of trying: effects of work environment and gender on post-adoption information technology use. MIS Quart 29:3

Ashurst C, Doherty NF, Peppard J (2008) Improving the impact of IT development projects: the benefits realization capability model. Eur J Inform Syst 17:352–370

Barker T, Frolick MN (2003) ERP implementation failure: a case study. Inform Syst Manag 20 (4):43–49

BCS (2004) The challenge of complex IT projects. British Computer Society, London

Berg N (2008) Secrets to a successful commercial software (COTS) implementation. iUniverse, Bloomington

Boudreau M-C, Robey D (2005) Enacting integrated information technology: a human agency perspective. Org Sci 16(1):3–18

Carr NG (2003) IT doesn't matter. Harvard Business Review, Boston, MA, pp 41–49

Chan YB, Gurbaxani V (2012) Information technology outsourcing, knowledge transfer, and firm productivity: an empirical analysis. MIS Quart 36:1043–1503

Clegg C, Shepherd C (2007) The biggest computer programme in the world ever: time for a change in mindset? J Inform Technol 22:212–221

Clegg C, Axtell C, Damodaran L, Farbey B, Hull R, Lloyd-Jones R, Nicholls J, Sell R, Tomlinson C (1997) Information technology: a study of performance and the role of human and organizational factors. Ergonomics 40(9):851–871

Currie WL (2011) Institutional theory of information technology. The Oxford Handbook of Management Information Systems: Critical Perspectives and New Directions 137

de Vries J, Boonstra A (2012) The influence of ERP implementation on the division of power at the production-sales interface. Int J Operat Prod Manag 32(10):1178–1198

DeSanctis G, Poole MS (1994) Capturing the complexity in advanced technology use: adaptive structuration theory. Organization 5(2):121–147

Doherty NF (2014) The role of socio-technical principles in leveraging meaningful benefits from IT investments. Applied Ergonomics 45(2):181–187

Doherty NF, Coombs CR, Loan Clarke J (2006) A re-conceptualization of the interpretive flexibility of information technologies: redressing the balance between the social and the technical. Eur J Inform Syst 15(6):569–582

Doherty NF, Champion D, Wang L (2010) A holistic approach to understanding the changing nature of organisational structure. Inform Technol People 23(2):116–135

Doherty NF, Ashurst C, Peppard J (2012) Factors affecting the successful realisation of benefits from systems development projects: findings from three case studies. J Inform Technol 27 (1):1–16

Dorgan SJ, Dowdy JJ (2004) When IT lifts productivity. McKinsey Quart 4:13–15

Eason K (1988) Information technology and organizational change. Taylor & Francis, London

Eason KD (2001) The process of introducing information technology. Behav Inform Technol 20 (5):323–328

Enders A, König A, Hungenberg H, Engelbertz T (2009) Towards an integrated perspective of strategy: the value-process framework. J Strategy Manag 2(1):76–96

Esteves J (2009) A benefits realisation road-map framework for ERP usage in small and medium-sized enterprises. J Enterprise Inform Manag 22(1/2):25–35

Fearon C, Manship S, McLaughlin H (2013) Making the case for "techno-change alignment": a processual approach for understanding technology-enabled organisational change. Eur Bus Rev 25(2):147–162

Fortune J, Peters G (2005) Information systems – achieving success by avoiding failure. Wiley, New York, NY

Gilbert AH, Pick RA, Ward SG (2012) Does 'IT doesn't matter' matter?: a study of innovation and information systems issues. Rev Bus Inform Syst 16(4):177–186

Grint K, Woolgar S (1997) The machine at work. Polity Press, Cambridge

Hanseth O, Jacucci E, Grisot M, Aanestad M (2006) Reflexive standardization: side effects and complexity in standard making. MIS Quart 30:563–581

Hughes A, Scott Morton MS (2006) The transforming power of complementary assets. MIT Sloan Manag Rev (Summer) 50e58

Jasperson JS, Carter PE, Zmud RW (2005) A comprehensive conceptualization of post-adoptive behaviors associated with information technology enabled work systems. MIS Quart 29 (3):525–557

Koch C (2001) BPR and ERP: realising a vision of process with IT. Bus Proc Manag J 7 (3):258–265

Leonardi PM (2007) Activating the informational capabilities of information technology for organizational change. Org Sci 18(5):813–831

Levinson M (2009) Recession causes rising IT project failure rates. CIO Mag 18 June 2009

Markus ML (2004) Techno-change management: using IT to drive organizational change. J Inform Technol 19(1):4–20

Marler JH, Lian X (2012) Information technology change, work complexity and service jobs: a contingent perspective. New Technol Work Employ 27(2):133–146

Martinsons M, Chong P (1999) The influence of human factors and specialist involvement on information systems success. Human Relat 52(1):123–152

Melville N, Kraemer K, Gurbaxani V (2004) Review: IT & organisational performance: an integrative model of IT business value. MIS Quart 28(2):283–322

Murphy GD, Artemis C, Unsworth K (2012) Differential effects of ERP systems on user outcomes—a longitudinal investigation. New Technol Work Emp 27(2):147–162

Nevo S, Wade MR (2010) The formation and value of IT-enabled resources: antecedents and consequences of synergistic relationships. MIS Quart 34(1):163–183

Orlikowski WJ (1996) Improvising organizational transformation over time: a situated change perspective. Inform Syst Res 7(1):63–92

Orlikowski WJ (2010) The socio-materiality of organisational life: considering technology in management research. Cambridge J Econ 34(1):125–141

Peppard J, Ward J (2005) Unlocking sustained business value from IT investments. California Management Review, Fall, pp 52–69

Peppard J, Ward J, Daniel E (2007) Managing the realization of business benefits from IT investments. MIS Quart Execut 6:1–11

Powell TC, Dent-Micallef A (1997) IT as competitive advantage: the role of human, business, & technology resources. Strategic Manag J 18(5):375–405

Ravichandran T, Liu Y (2011) Environmental factors, managerial processes, and information technology investment strategies. Decision Sci 42(3):537–574

Ravichandran T, Han S, Hasan I (2009) Effects of institutional pressures on information technology investments: an empirical investigation. IEEE Trans Eng Manag 56(4):677–691

Robey D, Boudreau M-C (1999) Accounting for the contradictory organizational consequences of information technology: theoretical directions and methodological implications. Inform Syst Res 10(2):176–185

Sauer C. Davis GB (2010) Information systems failure. Encyclopedia of Library and Information Sciences 3rd edn 1(1): 2643–2652

Shpilberg D, Berez S, Puryear R, Shah S (2007) Avoiding the alignment trap in information technology. MIT Sloan Manag Rev 49(1):51–58

Staehr L (2010) Understanding the role of managerial agency in achieving business benefits from ERP systems. Inform Syst J 20(3):213–238

Thatcher ME, Pingry DE (2007) Modeling the IT value paradox. Commun ACM 50(8):41–45

Ward J, Elvin R (1999) A new framework for managing IT-enabled business change. Inform Syst J 9(3):197–222

Waring T, Skoumpopoulou D (2012) An enterprise resource planning system innovation and its influence on organisational culture: a case study in higher education. Prometheus 30 (4):427–447

Wyld DC (2009) The utility of cloud computing as a new pricing-and consumption-model for information technology. Int J Database Manag Syst 1(1):1–20

Chapter 2
Transforming Field and Service Operations with Automation

Gilbert Owusu and Paul O'Brien

Abstract Severe cost pressures, attractive new markets and accelerating new product introductions have substantially increased the complexity of transforming service and field operations. Automating service and field operations offer a tremendous opportunity for achieving improvements in efficiency, cost savings and service delivery. At the heart of automating service and field operations is the efficient management of a company's resources. However, automating the decision-making process of addressing the imbalance between the supply and demand sides of the service provisioning and delivery has been a challenge for field and service operators. In this chapter, we outline a framework for addressing the challenge of transforming service and field organisations with IT. In particular, the framework helps to identify the type of transformation required, particularly the types of IT capabilities appropriate for deployment in an organisation. The framework has been used in BT, and we provide a case study of how we used it in implementing a service production management capability for managing BT's field force.

2.1 Introduction

Service is a key differentiator in a competitive marketplace (Vandermerwe and Rada 1988). Delivering high levels of service, whilst also managing costs, introducing new innovative products and sustaining an existing portfolio, is a challenge facing most companies today. Service is central to retaining customers as well as growing revenue and market share.

G. Owusu (✉) • P. O'Brien
Research and Innovation, BT Technology, Services and Operations, Martlesham Heath, UK
e-mail: gilbert.owusu@bt.com

G. Owusu et al. (eds.), *Transforming Field and Service Operations*,
DOI 10.1007/978-3-642-44970-3_2, © Springer-Verlag Berlin Heidelberg 2013

Operational effectiveness is at the root of excellent service. Across a range of service industries, from Internet retailing companies such as Amazon[1] to large-scale distribution companies such as UPS,[2] the experience of the end customer depends on service operations that perform in a consistent and predictable way. Successful service operations combine systems, resources and processes seamlessly. They rely upon an ability to flex and coordinate their resources, most notably their workforce, to ensure they are at the right place at the right time with the right resources to deliver service to customers.

Transforming the management of service operations, particularly field operations, offers significant opportunities for cost reduction and service improvement. Balancing resource supply to customer demand is the goal, and the speed at which a company can effectively respond to any imbalance represents a point of market differentiation. On the demand side, customer demand for specific services may be uncertain and variable. Service performance may be volatile or highly dependent on external factors such as weather. On the supply side, a typical service operator has multiple resource types including people and assets (i.e. physical and consumables), each with different capabilities and varying availability, spread across multiple locations. Effectively managing such resources so they meet demand every single day is the challenge facing service organisations.

In recent years, concepts from manufacturing such as Enterprise Resource Planning (ERP) have been adopted in service industries under the banner of *service production management* (Johnston 2005; Voudouris et al. 2008). Whereas traditional ERP focused on management of materials in manufacturing, service production management applies the same concepts and technology to service industries and the management of people and assets. The strategic benefits of service production management have become ever more critical to cost-effective service delivery. Service production management links strategic (long horizon) and operational (job-based) capacity planning (see Fig. 2.1) and adopts a similar technology set employed in traditional ERP systems. This is a key enabler for realising proactive and agile operations which dynamically flex supply to demand. In order to maximise the efficiency and effectiveness of complex production lines, a production management solution must coordinate the end-to-end demand and supply chains.

An effective ERP solution provides a distinct competitive advantage to a company. Wernerfelt's 'A Resource-Based View of the Firm' (1984) asserts that companies should focus on resources rather than products as a determiner of competitive advantage. Subsequent papers (see Fosser et al. 2008) emphasise the opportunity ERP implementations offer to support this view. This chapter outlines the challenges in implementing a service production management solution in a company and proposes two frameworks for mitigating risks associated with such implementations. A case study realising a service production management solution in BT is then presented.

---

[1] http://www.amazon.com

[2] http://www.ups.com

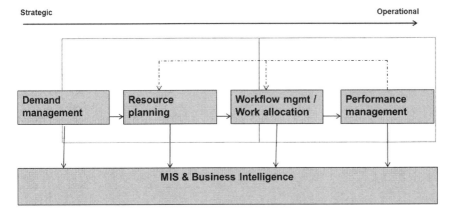

Fig. 2.1 Linking strategic with operational planning

2.2 Service Production Management: Challenges in Realising an ERP for Service Industries

2.2.1 Background

There are a growing number of companies embarking on service production management transformation programmes underpinned by production management technologies (PPF 2013). Despite the benefits that have been realised from these transformation programmes, the technologies that have been introduced have not been fully utilised leading to additional structural costs, slowness in responding to customer requests and potential loss of revenue opportunities (Davenport 1998).

Service production management transformation projects, as with many ERP implementations, are complex. Typically ERP implementations last between 6 months and 2 years costing on average \$1 M (2007) (Aloini et al. 2007); by their very nature they span business units, incorporate several component systems and involve significant organisational and operational change. Subsequently they hold significant risk. Many ERP implementations are unsuccessful. Aloini et al. (2007) cites ERP case studies where 90 % of SAP R/3 ERP projects ran late, another where 34 % of 7,400 IT projects were late or over-budget and only 24 % completed on time and on budget. Other papers cite a study where utility companies achieved less than 50 % value from an ERP implementation (Fosser et al. 2008). Mitigating such risks is central to realising a successful implementation. Davenport (1998) succinctly summarises the problem with many ERP implementations:

> Companies fail to reconcile the technological imperatives of the enterprise system with the business needs of the enterprise itself... An enterprise system, by its very nature, imposes its own logic on a company's strategy, organisation, and culture.

Table 2.1 Clustering of ranked critical success factors in ERP implementations from Somers and Nelson (2001)

Rank	Critical Success Factor	Man. of Change	Introduction of Technology
1st	Top management support		
2nd	Project team competence	✓	
3rd	Interdepartmental Co-operation	✓	
4th	Clear goals and objectives	✓	
5th	Project management	✓	
6th	Interdepartmental Communications	✓	
7th	Management of expectations	✓	
8th	Project champion	✓	
9th	Vendor Support		✓
10th	Careful package selection		✓
11th	Data analysis & conversion		✓
12th	Dedicated resources	✓	
13th	Steering Committee	✓	
14th	User Training	✓	
15th	Education on new Business Processes	✓	
16th	BPR	✓	
17th	Minimal Customisation		✓
18th	Architectural choices		✓
19th	Change Management	✓	
20th	Vendor Partnership		✓
21st	Vendor tools		✓
22nd	Use of consultants		✓

Somers and Nelson (2001) identify critical risk factors (CSF) associated with ERP implementation following a review of US industry implementations. The top factors (see Table 2.1 below) focus predominantly on the management of a transformation programme rather than technology selection. We can cluster these CSFs into those related to the management of change and those related to the introduction of a new technology.

Change occurs at different organisational and conceptual levels. In the next section, we briefly review the literature on the need for change, the models for change and institutionalising change with IT.

2.2.2 Management of Change

Management of change provides the framework for guiding change since the introduction of any new IT system will, of necessity, involve managing some degree of change to structures, processes, practices and often culture.

Johnston and Clark (2008, p. 3) observe that service operations management is concerned with delivering service to customers or users of the service. They contend that service operations management involves understanding customer needs and managing the processes that deliver services so as to meet any stated objectives. Oftentimes, managing the processes that deliver services requires changes to be made in the organisation. The literature is littered with many motivations for organisational change. Leana and Barry (2000) note that organisations pursue change to enhance their competitive positions and their adaptability in volatile markets. In particular, they observe that the reasons for organisational change may be motivated by one of the following external forces: *adaptability*, *cost containment*, *impatient capital markets*, *control* and *competitive advantage*. Capra (2003) notes that organisations need to undergo fundamental changes, both in order to adapt to the new business environment and to become ecologically sustainable.

There are several models for managing change. Morgan (1986/1997) argues that all theories (or models) of organisation and management are based on implicit images or metaphors which help in reasoning about organisations. The definitions of these models are generally drawn from three main disciplines—*behavioural science*, *psychoanalysis* and *systems thinking*. Most of the well-cited models combine themes from these disciplines. For example, Bolman and Deal's (2003) reframing approach is a comprehensive framework for analysing transformation programmes in organisations. They present four frames—symbolic, political, human resource and structural frames as lens for analysing organisations. Bolman and Deal (2003) describe the symbolic frame as a lens through which humans make sense of the messy, ambiguous world in which they live. They see meaning, belief and faith as the central concerns of the symbolic frame. Bolman and Deal (2003) view symbols as a way to resolve confusion, increase predictability and provide direction. However, with the political frame, organisations can be viewed as coalitions. They note that there are enduring differences among individuals and groups, thus making conflict endemic in organisations. This is amplified by finite nature of resources and the question of how they are allocated. Bolman and Deal (2003) argue that conflicts can be resolved by bargaining and negotiations. On the other hand, the structural frame focuses on the social architecture of work. It seeks to answer the key questions on how work is divided and coordinated. Mechanisms of coordination are either vertical or horizontal. One of the guiding principles of the structural frame is to design organisations so as to achieve efficiency and optimality. Finally, the human resource frame highlights relationships between people and organisations. They note that organisations exist to serve human needs, and a poor fit between people and organisations leads to poor performance. Aligning the needs of people and organisations requires an understanding of each group.

Tichy (1983) proposes the 'TPC' (i.e. technical, political and cultural) theory. He views managing change as involving making technical, political and cultural decisions about desired new organisational states; weighing the trade-offs; and then acting upon them. Other less theoretical models include Kotter (2007) and Kotter and Cohen's (2002) eight steps to transforming an organisation. The steps are:

1. Establishing a sense of urgency
2. Forming a powerful guiding coalition

3. Creating a vision
4. Communicating the vision
5. Empowering others to act on the vision
6. Planning and creating short-term wins
7. Consolidating improvements and producing still more changes, i.e. do not let up
8. Institutionalising new approaches or making the change stick

The introduction of any service production management system for transforming an organisation will impact both the instigators of the change and the recipients of the change. This clearly gives rise to different forms of resistance that any skilful change agent will need to understand and address. Drucker (1954) argues that the major hindrance to organisational growth (or maturity) is the inability of managers to change their attitudes and behaviour as rapidly as the organisations require. Armenakis et al. (1999) note two reasons why change efforts fail to become institutionalised—(a) impatience and assumption that successful change introduction and implementation guarantee institutionalisation and (b) the neglect of seeing change through to institutionalisation. Institutionalisation is about adopting a new mind-set.

Armenakis et al. (1999) observe that the commitment of stakeholders plays a major role in change initiatives. They view resistance to change as the same as commitment to the current state. As noted earlier IT projects are more than technical artefacts—there is a human element. The quote below from the Times Higher Education Supplement[3] emphasises this point:

> Recent research shows that about 80 % of IT projects fail to deliver stated business benefits because the "human dimension" has not been managed.

Clearly, more time should be devoted to creating the vision of transformation programmes, communicating them and empowering others to act accordingly—this requires a bottom-up approach and would engender commitment from all stakeholders.

2.2.3   Introduction of New Technology

Production management transformation is often accompanied by the introduction of new technologies for the management and operation of resources. Automated production management provides a system for optimising resource utilisation. In order for a system to optimise resource utilisation, it needs to know the potential demand from customers, the availability and capabilities of the resources and any business objectives that may govern quality of service. Typical applications of production management include demand forecasting, resource planning and scheduling, capacity reservations and appointing and revenue management.

---

[3] 18.02.00 Times Higher Education Supplement.

Most IT transformation projects adopt a mechanistic approach which focuses predominately on realising quick wins. However any quick win is short-lived if there are no corresponding political and human resource changes. Political and human resource changes do take time—they are for the long term. Participation and involvement of all the change recipients in requirements capturing phase, though desirable, take a long time and risk jeopardising the timely delivery of the IT system. Though a top-down-driven approach is required to provide the context for the transformation programme, not winning the hearts and minds of the stakeholder (i.e. a bottom-up approach) will lead to projects failing to deliver intended outcomes. Armenakis et al. (1999) highlight lack of time as the main reason for the failures of organisation change programmes—an attribute that tends to be lacking in most change programmes. There is an illusion of speed at the expense of producing a satisfying result, which is to change behaviours. Oftentimes there is no room for delays or feedback loops—two important factors in ensuring that a new way of thinking is put in place.

New IT systems will introduce frameworks for codifying domain knowledge and automating processes and practices. IT represents only one contributing factor to realising the full competitive advantage of a production management implementation. Significant competitive advantage lies in the *process* of realising ERP and production management solutions in a company as well as the IT implementation itself. As Fosser et al. (2008) succinctly put it:

> an ERP system alone does not create a sustainable competitive advantage ... managers can initiate processes based on the output of the ERP-system that can result in a competitive advantage ... Managers can foster an awareness of the creation, distribution and usage of this knowledge.

Established off-the-shelf IT products invariably have an embedded view of organisational design, business processes, data structures and user needs, all of which may require customisation if they are misaligned with the requirements for a particular implementation. Furthermore, customising a generic off-the-shelf ERP product, sometimes referred to as 'vanilla implementations' (Fosser et al. 2008), oversimplifies and avoids the key challenges in realising ERP. This experience is mirrored in other domains where off-the-shelf tools are used for realising operational support systems and have proved challenging; with 'vanilla implementations' significant customisation is required which subsequently fails to deliver the value originally anticipated (Owusu et al. 2008). This highlights the danger of vendor lock-in, where an organisation's processes and principles are locked into a specific software solution and a specific vendor roadmap (Kumar et al. 2003). Van Stijn and Wensley (2005) observe that:

> ...the notion of standard [ERP] templates is in some sense incoherent, since best practices are contextualized and we have to recognize that such practices will be interpreted or reinterpreted when they become part of and are enacted in the organization. The situations in which the practices exist or should come to exist are considered to be unique and that makes simply imitating them rather impossible. Further, once instantiated particular ERP land practices are not necessarily "best" for a particular organization

Researchers in management of change, particularly those with the challenge of IT transformation programmes, are faced with questions of how to ensure that such transformation programmes are successful and how to best align the intended organisational structure with the deployed IT system with planned processes resulting in the envisaged service. The starting point for us to address this challenge requires that we understand the mechanisms for transforming service and field operations with automation. In the next section, we outline two frameworks that provide the steps to de-risk the implementation of an IT-enabled transformation programme.

2.3   Realising a Successful Service Production Management Implementation

Two frameworks are proposed for addressing the risks associated with realising a successful service production management solution. In response to the issue of managing change, the 4 Cs maturity framework helps an organisation categorise service production management maturity and so understand the type of transformation required. In order to improve the likelihood of successful technology deployment in support of such transformations, an innovation-driven development methodology is proposed.

2.3.1   Maturity Framework: The 4 Cs

In light of the aforementioned issues to transform organisations using service production management technologies, we propose a maturity framework (4 Cs) to characterise different stages in the introduction of production management into a company. Different parts of the same organisation can be at different stages in this framework. The framework helps identify the type of transformation required, particularly the types of IT technology appropriate for deployment in an organisation (Fig. 2.2).

The 4Cs framework has been used to transform the field and desk-based teams in BT (Owusu et al. 2006; Voudouris et al. 2008). Each stage characterises different levels of sophistication in managing and forecasting demand, collaborating across value chains, proactivity in planning resources and agility when faced with perturbations in normal operations.

Complex characterises a highly manual approach to service resource management with ad hoc practices adopted across different business units, each responding in a suboptimal reactive way to incoming demand. This situation often occurs with highly decentralised organisations or often follows merger or acquisition activities resulting in service planning operating over short time frames.

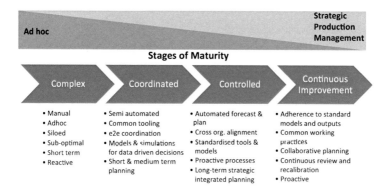

Fig. 2.2 The 4C framework

Coordinated organisations have a common approach to resource management. They have common IT tools and support within distinct business units which help enforce a coordinated approach to resource management. However the automation remains silted limiting the benefits from standards and automation. In coordinated organisations service planning operates at both medium- and short-term time frames.

Controlled enforces a common methodology to production management end to end across an organisation and extending into its demand and supply base. It introduces a more tightly integrated approach across an organisation, linking the market-facing systems such as CRM capturing demand through to the delivery units' managing service resources. A consensus will be reached between the demand and supply side of the business based on a data-driven analysis. This capability supports a more strategic outlook on resourcing and encourages a more proactive approach to resourcing. It involves employing a common data standard to ensure interoperability across business units, common tooling and models and something on agility.

Continuous improvement represents the final stage of maturity. Such organisations will have service production planning embedded in their culture. No organisation stands still. There will be continuous change in the marketplace, portfolio, governance, technology, regulation, skills, ownership, etc. In order to ensure such organisations sustain competitive advantage through effective service production management, it needs to embrace a mind-set where change is the norm, and it needs to pursue improvements continuously. Many management philosophies support this approach, the most common being TQM [Total Quality Management (Ishikawa 1985)].

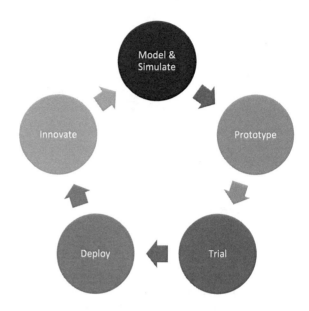

2.3.2   *Innovation-Driven Development*

The successful implementation of this framework has been underpinned by rigorous optimisation models and people engagement programmes. The success (i.e. adoption) of any IT transformation programme is in part a function of the quality of service being provided by the system (Fig. 2.3).

Transitioning between stages in the 4 Cs model is best achieved by using an agile development methodology. This approach has many advantages:

- Engaging users in trials helps to reduce risk associated with user reluctance to embrace transformational change by encouraging their participation in any transformation programme. It also helps engender the sense of ownership and engagement which creates advocates for the transformation across the organisation. Modelling and simulation helps to convince stakeholders of the usefulness of a system. By engaging end users in prototyping of models, they can quickly identify with what is being developed and will be inclined to use the system.
- Prototyping provides the environment for scenarios to be tested before committing to implementations. It allows a fail-fast mentality and subsequently can foster managed risk taking and innovation.
- Modelling and simulation assesses the viability of transformation programmes. It recognises that all operational change is complex, requiring the modelling and simulation of requirements and solutions. Simulation models generate additional information and insight about the challenges that require solutions in a transformation programme. They provide supportive change agents that are data driven that help convince sceptical users and can help flag issues prior to expensive

pilots and trials. IT projects initiate organisational change. In order to manage these changes effectively, the expectations and interests of the all the stakeholders can be managed properly. Successful change is about winning the minds and hearts of the people the change is intended for. Visual modelling and simulation will provide a collaborative framework to engage stakeholders from sales, finance, operations and strategy teams. The different stakeholders can provide their inputs into the development process and scope the projects. Visual modelling and simulation is a key enabler for innovation. By collaboratively working on the simulation models, the stakeholders can test assumptions and try out 'what-if' scenarios. From a finance viewpoint, affordability questions can be answered by the model. Operations will be to assess key performance indicators in the light of changes in demand and resource levels. Used properly, simulation and modelling will enable a thorough walkthrough of the to-be processes which will underpin the transformation programmes.

2.4   Case Study: BT's Optimisation and Planning for Field Engineers

BT sells products and services to consumers, small- and medium-sized enterprises, large corporates and the public sector. Approximately 23,000 field engineers offer a professional, coordinated and efficient field force to deliver communication services across the UK and globally. Field Force Optimisation Suite (aka FOS) was developed in-house by researchers in BT's technology, service and operations team. BT's challenge was to create an automated resource management system to optimise allocation of field resources to improve the quality of service. BT has a large workforce that serves geographically dispersed and diverse customers including businesses, ISPs and consumers. BT's existing systems and processes would be classed as 'coordinated' using the maturity framework (see Fig. 2.2). It had a largely common IT toolset across different business units which tended to be siloed organisationally and functionally.

The transformation programme was challenged to extend existing capabilities with both increased automation of resource management and delivery of a step change in integrated planning where a strategic view of demand and supply links into operational resourcing.

FOS was the chosen technology. It gives resource managers an overview of the work they are expected to complete in a specific time frame and the resources available to complete it. FieldForecast, the forecasting component, uses historical work volume statistics to forecast demand for different products and geographical areas over different time frames. The FOS staff management tool, FieldPeople, maintains employee data. It stores the field teams' preferred work locations, attendance patterns and skills. FieldPlan produces capacity plans based on forecasts, availability of people and business priorities. These components combine to allocate the right engineer to the right job at the right time.

The deployment of FOS followed the innovation lifecycle outlined above. Modelling and simulating different strategies for managing resources was an invaluable tool for de-risking development and trials, as well as providing a tool for agreeing requirements and limitation for the target system. A trial with a small team of users covering a limited geography was a particularly effective way of securing user engagement and advocacy and resolving teething problems with the systems. A common refrain was 'We are flagging up issues well ahead of time there are no surprises anymore' (PPF 2013).

2.5 Control of Workflow

The resulting system delivered significant improvements in automation, the quality of resource plans and overall transparency of field operations. A centralised approach to operational data meant planning was much less manual, and more than 80 % of the work scheduled was now allocated automatically. This meant that BT's resource managers now had more time to focus on short notice issues and opportunities. The FieldReserve system allowed resource managers to change the availability of appointment slots that the call centre can offer, where engineers need to make on-site visits. For example, they can remove slots in response to a sudden increase in demand, such as adverse weather conditions, and make sure that engineers are neither under- or over-allocated.

The FOS-enabled transformation project has significantly changed the way BT offers appointments to its customers. A web monitor gives a real-time view of appointment books with half-hour updates. Agents can see where there are gaps. Engineers with surplus hours can then be sent within their geographical areas of preference to an area where there is a potential backlog of need for their particular skill set. Reduced travel time due to better planning means more time on site with customers and more chance to complete the jobs while they are there. Because of this optimisation, the number of escalations—where engineers do not meet their appointment at the promised time—has been reduced from 15 % to 5 %. This transformation project has led to a 15 % improvement in cycle time, allowing BT to offer customers a shorter time window for visits, where access is required to business premises or homes.

2.6 Increased Efficiency Through Simplicity

The quality of the resource plans generated by the new system has reduced the complexity of resource planning process. It automatically analyses and processes large amounts of information and variables to optimise resource deployment. This has resulted in a closer match of supply and demand. Since the introduction of the new FOS system, more work is being covered by less people. The introduction of centralising planning with the new systems has reduced manual intervention from 31 % to 18 % plus c30 % improved productivity for resource planners.

Improvements in service levels have been seen due to more effective management of resources for service delivery. The deployed system has seen an 8–14 % improvement in the quality of service with a significant improvement in the number of technicians completing jobs first thing in the morning. Additional benefits of better plans have been a 20 % reduction in vehicle miles travelled by engineering fleet through the optimisation of job allocation. One of the end users summarises the benefits as follows:

'Before we were just focused on tomorrow, now we can plan further ahead, intelligently working with the information we have to make the best of our resources. We can keep up-to-date on the person's skills and how they change'. FOS Systems Subject Matter Expert (PPF 2013)

2.7   Conclusions

Service provides competitive advantage to any business or organisation. The successful realisation of service production management is a key component in delivering that competitive advantage. This chapter outlines the challenges associated with the successful realisation of a service production management capability. There are two types of risks in such an endeavour, the management of change and the introduction of new technology. In order to mitigate these risks, it proposes the 4Cs maturity framework and the innovation-driven development methodology. We presented a BT case study which describes the business impact following the introduction of a service production management solution for managing BTs field force. One of the key lessons we have learnt is to exploit the use of simulation models to garner stakeholder support when using IT transformation programmes to institutionalise change. We learnt very early in the programme to understand the motivations for change. This enabled us to effectively exploit capabilities in technology research, service management, change management and system engineering when developing and operationalising FOS. We also learnt to pay attention to end user requirements and used rapid prototyping to refine requirements.

References

Aloini D, Dulmin R, Mininno V (2007) Risk management in ERP project introduction. Rev Lit Inform Manag 44(2007):547–567

Armenakis AA, Harris SG, Field HS (1999) Making change permanent: a model for institutionalizing change interventions. Res Org Change Dev 12:97–128

Bolman L, Deal T (2003) Reframing organizations: artistry, choice and leadership, 3rd edn. Jossey-Bass, San Francisco

Capra F (2003) Life and leadership in organisations. In: Capra F (ed) The hidden connections. Flamingo, London, pp 85–112

Davenport TH (1998) Putting the enterprise into the enterprise system. Harv Bus Rev 76:121–131

Drucker PF (1954) The practice of management. Harper and Row, New York

Fosser E, Leister OH, Moe CE, Newman M (2008) Organisations and vanilla software: what do we know about Erp systems and competitive advantage? http://is2.lse.ac.uk/asp/aspecis/20080211.pdf

Ishikawa K (1985) What is total quality control? The Japanese way. Prentice-Hall, Upper Saddle River, NJ

Johnston R (2005) Service operations management: from the roots up. Int J Operat Prod Manag 25 (12):1298–1308

Johnston R, Clark G (2008) Service operations management: improving service delivery. Prentice Hall, Upper Saddle River, NJ

Kotter JP (2007) Leading change: why transformation efforts fail. Harv Bus Rev 85:96–103

Kotter J, Cohen SD (2002) Introduction: the heart of change. In: Kotter J, Cohen SD (eds) The heart of change. Harvard Business School Press, Boston, MA, pp 1–14

Kumar V, Maheshwari B, Kumar U (2003) An investigation of critical management issues in ERP implementation: empirical evidence from Canadian organizations. Technovation 23 (10):793–807

Leana C, Barry B (2000) Stability and change as simultaneous experiences in organizational life. Acad Manag Rev 25(4):753–759

Morgan G (1986/1997) Images of organisation. Sage, New York

Owusu G, Kern M, Voudouris C, Garyfalos A, Anim-Ansah G, Virginas B (2006) On optimising resource planning in BT plc with FOS. In: Proceedings of the international conference on service systems and service management, Troyes, pp 541–546

Owusu G, Anim-Ansah G, Kern M (2008) Strategic resource planning. In: Voudouris C, Owusu G, Dorne R, Lesaint D (eds) Service chain management: technology innovation for service business. Springer, Germany

PPF (2013) Professional Planning Forum Best Practice Guide 2013. www.planningforum.co.uk

Somers TM, Nelson K (2001): The impact of critical success factors across the stages of enterprise resource planning implementation. In: Proceedings of 34th Hawaii international conference on systems sciences (HICSS-34), January 3–4, Maui, Hawaii

Tichy N (1983) The essentials of strategic change management. J Bus Strategy 3:55–67

Vandermerwe S, Rada J (1988) Servitization of business: adding value by adding services. Eur Manag J 6(4):314–324

Van Stijn E, Wensley A (2005) ERP's best practices and change: an organizational memory mismatch approach. In: Bartmann D, Rajola F, Kallinikos J, Avison D, Winter R, Ein-Dor P, Becker J, Bodendorf F, Weinhardt C (eds) Proceedings of the 13th European conference on information systems, Regensburg, Germany

Voudouris C, Owusu G, Dorne R, Lesaint D (2008) Service chain management. Springer, Heidelberg

Part II
Methods, Models and Enabling Technologies for Transforming Service and Field Operations

Chapter 3
Designing Effective Operations: Balancing Multiple Business Objectives Using Simulation Models

Stephen A. Cassidy and David C. Wynn

Abstract All businesses need to balance a number of different and often competing imperatives. For a business to be sustainable it must balance two prime objectives: to maintain cash flow today for survival and at the same time to evolve, to ensure its future. Supporting these prime objectives are second-order objectives such as the ability to maintain business stability under various conditions. An example is the maintenance of customer service levels in the event of unexpected surges in demand or problems in supply. Managing these parallel demands is very complex, involving decision-making in different areas of the organisation, the results of which interact with each other on a range of timescales. New modelling techniques give us the ability to simulate the business behaviours which result from these interactions. They provide important evidence-based insights which improve decision-making in complex organisations.

3.1 Introduction

Simultaneous business requirements place multiple demands on resources and involve conflicting priorities. These include balancing long-term and short-term commercial impacts and the rewards between different stakeholders: shareholders, customers and employees. Multiple requirements and demands create stresses in the organisation, and these absorb significant continuous management effort to resolve. New approaches are called for which enable organisations to improve their performance over a balance of business objectives, rather than suboptimally pursuing a set of individual targets. New techniques in organisational modelling and powerful computer simulations can be used to create significant insights into the operation of

S.A. Cassidy (✉)
Research and Innovation, BT Technology, Services and Operations, Martlesham Heath, UK

D.C. Wynn
Engineering Design Centre, University of Cambridge, Cambridge, UK

G. Owusu et al. (eds.), *Transforming Field and Service Operations*,
DOI 10.1007/978-3-642-44970-3_3, © Springer-Verlag Berlin Heidelberg 2013

complex organisations. They simplify the visualisation of complex behaviours and make business assumptions and their impacts explicit. In this way they build common ground for achieving better balanced business decisions.

This chapter discusses the issues above and describes how computer simulation has been used to study a business process. The example chosen is the delivery of a telecom service. The delivery of customer connections to data services has been modelled, encompassing its interactions with adjacent business areas to give a more organisation-wide view of the resulting business and customer performance. Simulations have been used to reveal the balances between resource costs of processes and levels of customer service, in the light of decision structures and information flows between different areas of the business. These enable more informed decisions over competing business requirements.

3.2 Different Approaches to the Complexity Problem

3.2.1 The Problem of Complexity

A competitive service offering means one that is priced competitively and also one provided with a competitive service performance level. The latter also implies an acceptable level of predictability, which is needed to secure customer confidence in future delivery and so maintain the customer base. This customer confidence also affects the level of take-up of new services, so affecting future growth. Also affecting future growth is the ability of the organisation to innovate and renew itself. This determines the ability to deliver new products to market and the achievement of new efficiencies.

The effect on customers of different service performance factors thus impacts not only short-term performance but also the long-term performance of a company. Performance impacts on brand and reputation can have very long-term effects, both through their effect on customers and also on investors and potential partners. Likewise, business performance in the future is governed by the decisions made in the immediate term, such as organisational, investment and resourcing decisions.

It can be seen that this web of interactions, both in the external market and internal to the organisation, leads to significant complexity. This complexity can take a lot of unravelling, and the results can involve a level of unpredictability. So how do we deal with this problem?

3.2.2 Current Organisational Approaches

Most often, organisations tackle the different business concerns by setting up distinct initiatives or management structures, each designed to ensure a focus on one aspect: be it cost, service levels, etc. Owing to this complexity of the external market and the internal organisation, it is often difficult to understand or predict the consequent interactions between these initiatives. There is a lack of an accepted common methodology or even language about how we understand complex, 'organic' organisations. In the absence of methodical approaches and the ability to simulate complex systems, the only possible approach has to be one which allows the different initiatives to compete with each other within the organisation, and the organisation itself becomes the model used to investigate the problem.

Working these out through the organisation itself can be expensive and ineffi- cient. In addition, the results are necessarily found after the fact and without an awareness of the near alternatives and their possible outcomes. An ability to 'experiment' on a model of the organisation, through simulation, enables a whole range of alternatives to be assessed efficiently. Ideally these simulations would help explore levels of possible risk and unpredictability and help understand patterns of behaviours in cases where absolute predictability might not be possible. This would give a more realistic view of the possible outcomes and lead to more enlightened business decisions.

3.2.3   A New Approach

This more successful approach develops ways to understand the organisation in a structured way, taking into account the fact that each of its internal parts and the various environmental (customer, market, competitive, technological) factors inter- act with each other. These interactions are asynchronous and decisions made in each part of the organisation depend on the information supplied to them by other parts of the business and from the external environment.

It is essential to understand the *nature* of the external environment. Different expectations of the stability of demand and the unpredictability of market change lead to very different organisational characteristics needed for successful behav- iour. Explicit characterisation of the expected levels of market stability and vola- tility is fundamental to designing a successful organisation. Mismatching the organisation to the environment can lead to an over-expensive market offering or conversely an inability to adapt to changing market conditions.

We have found it useful to apply the principles of *cybernetics* to understand the organisation (Weiner 1950). This treats a system as a set of interrelated parts, and each functions in the organisation as one of a set of control loops. Central to the cybernetic view is the concept of *variety* and the importance of matching the variety of the external environment with the *requisite level of variety* inside the organisa- tion (Ashby 1958).

This means that we must appropriately characterise the variety in the external market. Key elements of variety include variation in day-to-day demand, and the level to which it can be predicted, and also the expected rate of change of what the market might demand in the future, and its level of predictability. We have developed a way to decompose different aspects of variety or *volatility* in the environment.

A computer model has been developed based on the principles described. It views the organisation as a 'complete' set of interacting parts, where 'complete' means all those parts which have some bearing on the business performance in question. We have explored its response to different external demands, where these have been chosen to exemplify different types of variety according to our description. The responses have been explored in terms of a range of business measures—cost, resource use, customer service levels, time to recovery of normal performance, etc. At this stage we have not explored the ability of the organisation to change or innovate; this is the subject of continuing research.

The model has been used to study the effects of different decision mechanisms and the presence or absence of specific information flows between different decision points in the organisation, for example, whether resourcing decisions are taken in response to observed performance change, whether (and to what extent) historical data can be used to estimate future demand or whether other parts of the organisation have information which could give earlier insights on future market demand.

The extensive data from exploratory runs of the model have been visualised in ways which allow insights to be gained quickly. Examples include revealing counterbalancing factors which indicate fundamental trade-offs between performance factors and showing the spread of possible outcomes, building insights into business risks. In turn, these visualisations also enable business experts to express further insights and so provide useful inputs to the development of the modelling. The process of discussion makes any assumptions explicit, which is essential to any quality decision-making process.

The rest of the chapter follows the structure outlined, in the following format:

- Characterisation

  - Of the organisation—cybernetics
  - Of the environment—dimensions of variety

- The model

  - Constructing the model—methodology and structure
  - The effects of control loops and information flows in the organisation
  - Explorations of process robustness—characterising the limitations and balances between business priorities: customer service levels and cost-efficiency

- Insights generated

  - Balancing business performance metrics
  - The effects of enterprise data quality on performance—technical, organisational and human impacts and dependencies

- Recommendations

  - Practices
  - Decision-making in organisations

3.3   Characterisation of the Organisation and Its Environment

3.3.1   Characterising the Organisation: Cybernetics

A useful way to model a regulated system is to use the principles of cybernetics. This approach understands an organisation as a set of interacting parts and in which certain key concepts are contained.

Feedforward and Feedback. Perhaps the most basic principle of cybernetics states that a system delivers desired behaviour in an uncertain and changing environment through feedback and feedforward mechanisms. The system (here, organisation) has a set of inputs (in this case, demand and other information from the market) which are converted into a set of outputs (here, aspects of performance) using control parameters to adjust its behaviour. In feedback, the difference between aspects of the system's performance and their target values is measured, thereby creating information about the interaction between the system and environment. This information is processed and used to change the control parameters to bring the output closer to its desired state. In feedforward, the process is similar, but measurements are made at the system input (and within the system?). This offers the possibility of midstream course correction, in advance of problems being manifested at the output of the system.

Inertia and Delay Effects on System Stability. Most controlled systems display inertia, taking some time for a change in the control parameters to become visible in the output. Forms of inertia in a business process include the work that builds up in queues when insufficient resource is available, assembling information that the process uses, and the base of 'integrative' resources, such as infrastructure that must be maintained. If the target values or the inputs change, most inertial systems governed by feedback control will undergo oscillations before settling close to the desired value. A system which diverges from the desired value as time progresses, or which may unpredictably settle close to one of several different states, is said to be unstable. In general, the instability and oscillatory behaviour of a system are exacerbated by increased delays between cause and effect and by interacting control loops. In an organisation, regulatory processes with long lead times thus tend to cause excess oscillations. Oscillatory effects can be seen, for instance, when too much resource is allocated too late to deal with a temporary increase in workload, only to be left with excess capacity that must be paid for once the job is finished. Generally, the problems associated with instability can be minimised either by reducing lead times or by taking better account of the system dynamics in the control laws.

Principle of Requisite Variety. This general principle states that, to respond to a certain level of variety in its environmental conditions, a system must have requisite variety in its internal structure and/or control laws. The more variation or unpredictability in the demands from the environment, the greater the need for sophistication in the control laws and the information flows in the organisation for it to perform successfully. If no variation in demand for a service exists, no measures and balances are required in the organisation; a very simple process is sufficient, operating like a production line with defined work rates and steady flow. However, if the demand becomes more complex or uncertain, additional resource allocation loops and other controls are required to respond appropriately.

In summary, viewing an organisation as a cybernetic system highlights that each business unit and process plays a different regulatory role. These roles can be considered to construct a simulation of that organisation. The major regulation decisions (such as when and how much resource should be added) can be identified and included in the model. The broader context can be considered to identify what additional information could be used to assist these decisions. The impact of different control laws and information flows (feedback and feedforward) can be explored through simulation. This enables us to assess their values under different operating conditions. Tuning the behaviours of the organisation appropriately to the expected environmental characteristics is crucial to successful performance. It is therefore essential to characterise the expected environment in an equally explicit and systematic way.

3.3.2   Characterising the Environment: Dimensions of Variety

To characterise the operating environment, it is useful to consider two dimensions of uncertainty. The first is the level of unpredictability in the *quantity* of demand, where *what* is delivered is known. Within the context of delivering a familiar product, there is a spectrum of demand uncertainty. On the one hand, in a stable market, the statistics of historical demand can be used to forecast future demand. The other end of this scale represents a less stable environment. In this case it may not be possible to predict future demand, or it may be felt necessary to build the resilience to cope with unexpected events.

The second dimension of uncertainty is in *what* it is that is delivered, or how it is delivered. Environmental (market) changes may require the organisation to develop a new product or to transform the way a product is delivered. In this case the range of uncertainty is in the level of ambiguity in the definition of the new product or delivery mechanism. On the one hand we may be building to a well-understood requirement, or we may need to discover and adapt as we progress.

These two dimensions of uncertainty give us four regions as follows. See Fig. 3.1.

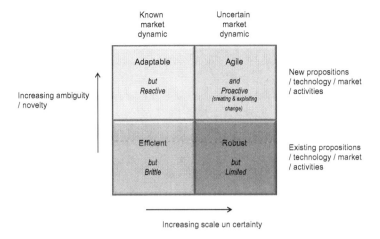

Fig. 3.1 Box model: dimensions of uncertainty in the environment

Efficient Operations. A system may be designed to operate efficiently under a given operating regime, where uncertainties exist but can be localised and quantified. For instance, although a service operation is usually subject to variable demand, historical data allows the numbers and types of customer request to be modelled using probability distributions. Using simulation, it is possible to tune the processes to handle that pattern of demand as efficiently as possible, to reach a desired trade-off between performance indicators such as lead time and cost. This requires that the process comprises fully determined elements with a high degree of standardisation. Resource allocation is highly optimised, and headroom is pared down to ensure efficiency. A highly tuned system like this is *brittle*, in the sense that if an unexpected demand does arise, performance is likely to suffer more significantly than a *robust* system.

Robust Operations. A process system may also be designed to cope with hitherto inexperienced variations, which occur when demand moves outside its normal window of operation. For instance, the expected pattern of demand might be disturbed by an unexpected and unplanned event, such as a significant increase in orders or a reduction in resourcing levels. Several strategies might be used to cope with the possibility of such events, including:

1. *Relax the performance targets.* Adding a few days buffer, although increasing average delivery time, could be outweighed by a reduction in delivery time uncertainty.
2. *Make the process less efficient.* Tuning a process system to be less efficient, for instance, by having extra resources on standby, allows more headroom to cope with unexpected needs.
3. *Make the organisation retune itself after a problem is spotted (feedback).* If a process system contains more 'intelligence', it may be able to identify changes in operating conditions and retune its own behaviour. For instance, consider an

unplanned increase in demand, which may create a *backlog* of work; if the backlog is monitored, the increase can be spotted and additional resource assigned to reduce it. On a higher level, if these adjustments repeatedly prove ineffective, the policy for allocating resource might be changed.

4. *Feed information between business units to anticipate the change in conditions* (*feedforward*). If a change in operating conditions can be anticipated, for instance, by the sales department who have some ability to foresee change in customer demand, appropriate actions can be put in place to compensate in good time, and major disturbances can be avoided.

At this level, the organisation is resilient, but additional capability is needed to give it the ability to *adapt* to changing requirements in the market.

Adaptive Organisation. An adaptive organisation is able to make decisions at all levels that consider issues that span multiple products and timeframe concerns in a portfolio. For instance, it may rationally decide to allocate more resource to improve support processes, recognising the ultimate impact on many primary processes and on specific product areas. An adaptive organisation can readily make trade-offs between short-term and long-term goals, for instance, by one process relinquishing resource now so that it can be used to improve performance later. It is able to rebalance the mix of what is done, rather than simply tuning the existing activities. It is also able to consider how it defines the market for its services through its own interactions with customers.

Agile Organisation. A fully agile organisation is able to redesign itself in response to major events that are outside prior experience and which fall outside the scope of the first three levels. Typically, this cannot be easily dealt with through a process-oriented analysis. Efforts towards an effective organisation might focus on how the team can dynamically create new ways of working appropriately to truly unexpected situations.

The analysis of adaptive and agile organisations is outside the scope of the work described here but is being considered in continuing research. We instead concentrate this analysis on efficiency and robustness.

3.4 The Model

3.4.1 Constructing the Model

Business processes were analysed for their response to these levels of uncertainty using discrete-event simulation. The model elements included proactive and reactive organisational functions, external events, internal 'models' (information stores, which are called on by processes and populated by others), information passing and logic functions. A Cambridge Advanced Modeller (CAM) toolbox was created to construct and simulate the models. The models are constructed through workshops with business areas, capturing the structure of business units, processes, information and job flows (Wynn et al. 2012).

The example chosen was the delivery of data circuits to end customers. This followed the taking of the order, its validation and the various installation and configuration stages, through to connection and closure of the order. The model encompassed a whole range of interacting business functions such as the processing of orders, the monitoring of waiting times and the resulting resource adjustments in the process, the frequency of the monitoring process which ensures sufficient network capacity and the financial and ordering processes for any new equipment needed. With these were the various associated decision processes, their characteristic timescales and information needed to trigger those decisions. The volume of demand and the variety in order complexity were taken into account.

The workshop participants were asked to estimate numeric data that would later be used to populate and calibrate the simulation. This included the lead time of each function as well as the frequency of certain operations. Some calibration data were obtained from company databases. Other information was estimated by the participants who were familiar with the process. A key element of numeric data is to identify the root causes of uncertainty in the duration of each job. In each case, workshop participants were asked to consider if variability could be attributed to properties of the work at hand. One property thus identified was the complexity of customer orders. A complex order would require more validation steps and take longer to design, than a straightforward 'off-the-shelf' installation. In the model, this parameter would be defined when each order was received and then influence the duration of several steps throughout the process.

Finally, the participants were asked to identify key performance indicators (KPIs) for the service provision and estimate expected values for each KPI; this information was later used to help calibrate the simulation. The resulting model, shown in Fig. 3.2, provides a simplified representation of the organisation, its main processes and the links between them.

3.4.2   Control Loops and Information Flows

Three main regulation loops were implemented in the model:

Resource regulation. The decision regarding how much resource to add to the process following an observed increase in congestion was implemented as a PD control law, tuned to find parameters that brought queue length under control quickly and with minimal oscillation following a disturbance.

Network capacity extension. The capacity available to support new installations is observed regularly; if it drops below a threshold, a defined 'increment' of new capacity is ordered. After some delay, this becomes available for use.

Demand regulation. Portfolio management is responsible for balancing volume by initiating a market campaign once the number of orders being placed falls below

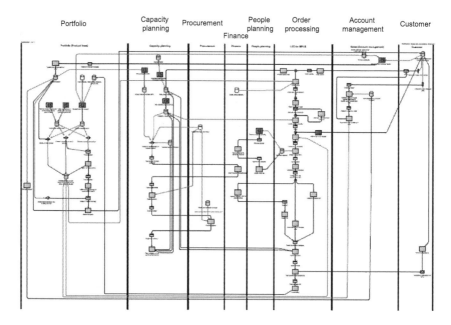

Fig. 3.2 Model of service operations for simulation: swim lanes delineate business units

a certain threshold. After some delay, this is assumed to cause a jump in demand, which then gradually tails off at a constant rate. In the model, this effect only comes into play during simulations with long timescales spanning several years.

These regulation loops span different functions in the organisation: the order process, resource management, capacity management and portfolio management.

The model enabled application of the conceptual framework described in earlier sections to explore opportunities for improvement in service provision. A set of analyses was run to consider the first three of the four levels of demand dynamic discussed above. The following section discusses the analysis undertaken to explore robust processes (the second level above), which is representative of the others.

3.4.3   Simulations to Explore Process Robustness

Analyses were carried out to find an efficient operating policy that is able to cope with unexpected steps in demand, i.e. robust to this kind of variation. The first step was to identify whether such a setup could be created through policy adjustments within the existing process structure. Five policy levers were identified, one example being running the capacity recovery process more frequently, another being the scale and speed of resource allocation following an increase in workload. These represent the control parameters of the process system.

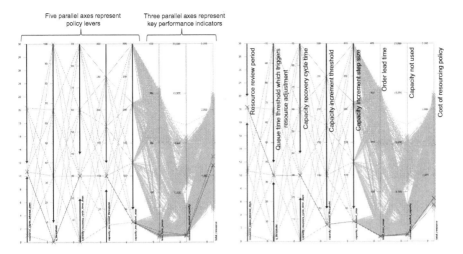

Fig. 3.3 Optimal policies for process operating under noisy demand (*left*) and noise combined with a step in demand (*right*)

A full-factorial simulation experiment was designed to evaluate different combinations of the five control parameters under each of the two operating conditions: the first being constant average, but noisy, demand and the second being the same but adding a step increase. All five control parameters were varied in four steps. For each configuration, 20 simulation runs were executed, allowing the effects of the noise to be indicated. For each of the two operating conditions, the results were analysed to identify the combination of the five levers that minimises customer response lead time without requiring very high resourcing levels. This analysis was performed graphically in the CAM software, using an interactive selector on the parallel co-ordinate plots shown in Fig. 3.3.

The original workshop participants worked with the model to explore the interrelations between control parameters and business performance and the sensitivities to different parameters and their combinations. This process of exploration generated important business insights as well as reinforcing certain intuitions. These have been used to inform a number of business policies beyond the scope of this chapter.

Comparison of the two plots shows that a different policy is required to give good performance in the case of unplanned disturbances (right plot) than for the case of noisy, yet predictable, demand (left plot). Comparing the policies depicted in the plots, the first and second policy levers (time between opportunities to adjust resource levels and the queuing time threshold at which more process resource is added) are set higher, and the third (cycle time of unused network capacity recovery process) is set lower in the right plot. In terms of KPIs resulting from these two ideal policies, it was found that the total resource requirement and wasted resource must be significantly higher to cope effectively with the step in demand (this is shown by the difference in scale between the labelled axes on the two plots).

These results can be explained as follows. A process system that is tuned to the limits of its performance has fewer built-in buffers so that when unexpected events occur, there is less slack to cope with them. In terms of the system dynamics, a process carefully tuned to respond quickly to small variations is less stable and tends to overshoot when large disturbances occur. In other words, the process as modelled can be either robust to disturbance or efficient under normal demand. Of course, the organisation would be interested in a process system that gives the best of both worlds—it would like to be both efficient and robust. The model helps determine the optimum balance point between these two for any given assumption about the nature of the environment.

These results help determine the performance limits of 'tuning' the existing process design. An alternative approach is suggested by the principle of requisite variety. This principle implies that a more complex environment can be dealt with by a more complex system. In this case, the principle suggests that additional information feeds and/or more complex regulation principles could help the organisation account efficiently for both variability and unexpected events in its operating environment. This could be achieved, for example, by a feedforward flow that predicted a disturbance. Such a prediction could be sent to the resourcing process by the sales business unit, because the latter is closely connected to the customer and thus has some capability to foresee changes in the pattern of demand.

To evaluate this proposed process improvement, the model was altered to include an extra information flow from sales into capacity planning, allowing an early warning of increases in demand to 'short-cut' the delays introduced by the long lead time resourcing process and thereby ensure that the process has sufficient capacity to deal with the disturbance when it occurs.

Assuming that the change in demand can be perfectly predicted and the exact amount of resource put in place, simulation revealed that the process is able to smooth out the 'bump' with no significant effect upon the main performance parameters. However, if the prediction is not perfect, the extra resource might be put in place too early or too late. A simulation experiment was therefore created to assess the impact of poor prediction fidelity on the performance of the most efficient design as determined in the model of the simpler organisation. The results confirm that perfect prediction of bumps in demand gives better process performance than poor prediction.

The dependence on performance of prediction quality is show in Fig. 3.4.

The cross hairs in each plot show the centre of the cluster of cost and delivery times of individual model runs (which, for clarity, are not themselves shown), where the only uncertainty is within the statistics of the variable demand, and for an organisation with no feedforward view of the future. In other words it is optimally tuned to react to its performance and demand as it happens. The clusters of points represent the statistical spread of model runs where feedforward demand prediction is used. In the left-hand plot, the prediction is perfect, and on the right the prediction is very poor; the middle plot represents an intermediate position.

This visualisation shows immediately that the result of good prediction is to bring the delivery time under control: reducing it and hugely reducing its spread.

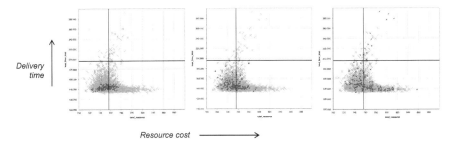

Fig. 3.4 Process performance effects for different prediction accuracies: high, medium, low

This is very beneficial for the customer experience. The average and spread in the cost are also reduced but not so dramatically. Looking at the sequence of plots, as the quality of prediction is reduced, the performance of both cost and delivery time deteriorates, and their spreads increase. Again, delivery time is more sensitive to this than the cost. There comes a point where bad prediction may be worse than no prediction at all, and the control is best left to the reactive feedback mechanisms in the organisation. However, at its best, feedforward delivers significant increases in customer satisfaction and cost control at the same time.

3.5   Insights

3.5.1   Balancing Performance Metrics

The simulations above show the effects in cost and service performance of different combinations of policies in the organisation. These include resourcing levels, and how quickly they can be adjusted, the frequency of capacity checking and upgrade, etc. The visualisations allow these 'policy levers' to be adjusted to guide the expert towards the most beneficial ways to combine these policies to generate the desired set of performance targets for the organisation. The graphs in Fig. 3.3 immediately show practical limits to these policies. For example, it may in theory be desirable to have a policy of very low resource margin and at the same time a very low and controlled delivery time target. The model shows the limits and trade-offs that are possible in practice, in other words where further pursuit of a particular set of targets could even be destructive to performance, there being no successful run of the process with that set of parameters in the simulation.

The model visualisations also provide further, broader levels of insights into the factors which determine the performance of the organisation. For example, looking at the plots in Fig. 3.4, it can be seen that reducing the accuracy of the feedforward information increases (modestly) the resource cost of the process, averaged over many runs, and more quickly results in increased delivery time and the uncertainty within it. We might ask what factors in the organisation influence the accuracy of the feedforward information. These factors include technical, organisational and human aspects and are explored in the next section.

3.5.2   The Effects of Enterprise Data Quality on Performance

Making informed decisions which tune a complex organisation to perform to a number of interacting and competing business metrics is a daunting task. The techniques described here provide tools to help make this task more tractable. Since they are built on data, they provide an evidential base for decisions. The simulations show the extent to which data quality and availability in different parts of the organisation limit the ability to optimise performance. Quality data is therefore a key enterprise resource, and the technical, organisational and human dependencies which lie behind data quality need to be considered.

3.5.2.1   Technical, Organisational and Human Aspects

Accuracy of feedforward information in an organisation is influenced by a number of factors. These include the more *technical* aspects of data accuracy and cleanliness and the quality of any algorithms which might operate on that data.

They also include *organisational* aspects, such as departmental boundaries which can either result in poor transfer of data across those boundaries or make the data less useful having crossed such a boundary. Each of the different parts of an organisation may have different perspectives on data—why it is gathered, its form, labelling and scope—driven by the different local purposes of the part of the organisation. If the data is not gathered with an enterprise-wide perspective, enterprise-wide benefits will be more difficult to achieve.

Finally, there are the *human* aspects. Data is either entered into systems directly by people, or in the case of automated data gathering, the data design, rather than its gathering, is a human activity. Accurate, unbiased and clean data is therefore fundamentally dependent upon the motivation of the employee and what they were hoping to achieve at the point of data design or gathering. Their purpose may have been driven by an enlightened view of the enterprise and its improved performance or possibly a more localised benefit in one area. At the extreme, local targets placed on individuals may lead to gaming or the seeking of ways to minimise the troublesome overhead of data entry.

For feedforward data to be maximally useful in improving organisational performance, the local motivations, organisational influences and technical aspects need to be carefully considered. The organisational modelling described here can therefore help identify critical points of intervention in any of these three aspects of the organisation.

3.6   Conclusions

Balancing several, often conflicting, business objectives in a complex organisation presents considerable challenges. Dealing with these challenges absorbs a great deal of management energy, involving multiple continuing interactions and negotiations between different stakeholders in the business. The full effects of decisions taken are often seen well after the event, which often give rise to the need for further course corrections. New techniques such as those described in this chapter offer the possibility to anticipate the results of interacting decisions and policies in an organisation, so adding a greater degree of insight into the range of possible actions and their consequences.

These conclusions suggest two principal recommendations for organisations to take advantage of these techniques: Firstly, to build practices which use models as a means to explore, clarify and articulate business issues; and secondly, to use the models to help different stakeholders in the business, understand the effects of their decisions on each other and on the broader business objectives. These models have the capability to provide enhanced levels of evidence into these discussions.

3.6.1   Practices

The modelling techniques described help build a practice in the organisation where interaction in workshops gathers information but also provides feedback into the development and structure of the model and a better understanding of the focus questions such models should answer. It helps clarify those questions and broadens explicit agreement on the assumptions being made. Future developments will broaden the set of business decisions that we will be able to inform using such models.

3.6.2   Organisational Decision-Making

Scientific approaches to decision-making, grounded in data, present new opportunities for understanding complex organisations. Dynamic models like the one described here enable the possibilities to be investigated with increasing thoroughness and the likely behaviour of an organisation under different environmental conditions simulated. Increased computing power means that the effects of many parameters can be explored in a reasonable time. This enables diverse but interdependent stakeholders, with different local aims, to understand the effects of their decisions on other parts of the organisation and its performance as a whole. It is well known that people, no matter how intelligent, struggle to control even simple systems under conditions of feedback and delay (Diehl and Sterman 1995).

Large service organisations deal with a very large number and variety of nested delays and multiple feedback mechanisms and daily allocate a resource of many thousands of employees. The use of these tools will improve the performance of the organisation across a balance of business objectives by promoting a wider shared understanding amongst diverse areas of the organisation and by bringing greater coherence to decision-making and co-ordination across the organisation.

References

Ashby WR (1958) Requisite variety and its implications for the control of complex systems. Cybernetica 1(2):83–99

Diehl A, Sterman JD (1995) Effects of feedback complexity on dynamic decision making. Org Behav Hum Decision Proc 62(2):198–215

Weiner N (1950) The human use of human beings: cybernetics and society. Houghton Mifflin, Oxford

Wynn DC, Cassidy SA, Clarkson PJ (2012) Design of robust service operations using cybernetic principles and simulation. In: Proceedings of the 12th International Design Conference, DESIGN

Chapter 4
System Dynamics Models of Field Force Operations

Kjeld Jensen, Michael Lyons, and Nicola Buckhurst

Abstract This chapter discusses an approach to modelling service operations appropriate to long-term strategic planning which incorporates the modelling of service performance as an integral feature of the methodology. The approach is based on the system dynamics technique, which emphasises dynamic complexity over detail complexity. We present the rationale behind the approach, with a key principle being that the combined effect of decisions, e.g. on resource deployment, tends to equalise tension across the organisation with 'tension' represented by equations relating performance, targets and prioritisations. We use a simple implementation of this approach to demonstrate the types of scenarios that we have explored with clients, followed by a description of how we have used a more comprehensive version, validated against historical data, to model the field operations of a major BT line of business, including a discussion of some of the challenges faced.

4.1 Introduction

Over the past few decades, both manufacturing productivity and the quality of manufactured goods have increased enormously. In contrast, service productivity has only shown small improvement, whilst quality has either shown no improvement or, in many areas, has declined (Oliva and Sterman 2001, 2010). Many factors contribute to this difference, but a number of authors have pointed out that because services are produced and consumed immediately, there is no finished goods inventory to cope with the inevitable variability in demand (Sasser 1976; Dietrich 2006). Hence, 'service providers are particularly vulnerable to imbalances between supply and demand' (Oliva and Sterman 2010).

K. Jensen (✉) • M. Lyons • N. Buckhurst
Research and Innovation, BT Technology, Services and Operations, Martlesham Heath, UK
e-mail: kjeld.jensen@bt.com

G. Owusu et al. (eds.), *Transforming Field and Service Operations*, 47
DOI 10.1007/978-3-642-44970-3_4, © Springer-Verlag Berlin Heidelberg 2013

Sasser (1976) argued that a service provider's ability to achieve a balance between supply and demand was a major determinant in its success. He proposed two basic strategies: 'chase demand', where the provider seeks to vary supply in step with changing demand, and 'level capacity', where capacity is based on average demand. Most service providers use a mixed strategy. Whilst it may be possible to define an average demand, there are usually significant short-term deviations from this average value. If delays in providing service are to be avoided, then some flexibility (e.g. overtime) is needed to cope with short periods of high demand. Furthermore, in the long term, the average demand is likely to increase or decrease.

Thus, in a service company like BT, operational managers, who are expected to deliver to tight timescales whilst controlling costs, need to understand the balance between service performance and costs. At an organisational level, there is a need for both tactical and strategic operational planning. Tactical planning focuses on timescales of days and weeks with a view to planning the deployment of resources to meet immediate demand; strategic planning is concerned with longer-term issues such as overall resourcing levels, performance metrics, targets, work prioritisation and the organisation's ability to respond to contingencies.

There is a need for modelling tools that support both tactical and strategic planning. In this chapter, our focus is towards the strategic end of this spectrum and we describe the development of dynamic models that help managers under-stand how demand, resource levels and performance are all interlinked. These models can be used to quantify trade-offs between costs and service performance. This, in turn, can be used to drive a service-performance-led approach to long-term planning in terms of resourcing levels and dynamic resource deployment policies.

The structure of the chapter is as follows: In the next section, we discuss the modelling of service performance and outline our modelling approach. We then use a fairly simple implementation of this approach to demonstrate the types of scenarios that can be analysed and illustrate the organisational behaviour encapsu-lated in a typical model. In the following section, we discuss a more complete implementation of the approach, developed to support planning of field operations in one of BT's major lines of business. Finally, to summarise the lessons learnt, we describe some of the challenges and limitations encountered.

4.2   Modelling Service Operations

4.2.1   The 'Hydraulics' Approach and Field Operations

Although we have developed the details of the approach, as described in this chapter, it does build significantly on the system dynamics literature, going right back to Forrester's earliest models of organisations, where differences between targets and actuals were used to drive adjustments of resource deployment (Sterman 2000 Chap. 15; Forrester 1960).

We frequently use a 'hydraulics' metaphor to describe this methodology due to its analogy with reservoirs, flows and pressures in fluid mechanics and will refer to it using that name in the following discussion.

We have used this 'hydraulics' approach to model a range of operational scenarios within our company including modelling of call centres and IT support teams. However, most of our modelling has been applied to field operations teams. These teams are faced with a number of workstreams including the provision of different types of services, as well as repair work. Each of these workstreams has a different target cycle time to meet.

In order to cope with variations in demand, a number of strategies may be employed. For example:

- Multi-skilling. This allows the staff to be moved between different workstreams thus maximising the utilisation of individuals.
- Overtime.
- Use of contractors.
- Manipulation of overhead time or 'shrinkage'. This comprises a number of necessary but non-productive activities such as annual and sick leave, team meetings and training. Whilst some of these activities are unavoidable, it is possible to defer others to address short-term peaks in demand.
- Defer own work. This is where the organisation postpones work that is not directly related to customer demand, e.g. routine maintenance and equipment upgrades. Like manipulating shrinkage, this is a short-term option. In the long term, the work has to be made up.

4.2.2   Dynamic vs. Detail Complexity

Field operations are by their nature complex. Senge (1990) distinguishes between complexity arising simply from there being many variables (*detail complexity*) and that arising from situations where cause and effect are interlinked so the effect, over time, of an intervention is not obvious (*dynamic complexity*). Field operations in service organisations in most cases would exhibit both flavours of complexity. This poses a challenge to the planning and management of such operations, presenting a requirement for appropriate planning and modelling tools with an adequate level of complexity and sophistication.

There are many modelling approaches that support tactical planning, such as the suites of modelling and planning tools presented elsewhere in this book. Amongst modelling and simulation methods, discrete event simulation (DES) is the method most often applied to tactical operational planning where detail complexity and stochastic variability are important (Chahal and Eldabi 2000, Brailsford et al. 2010, Morecroft and Robertson 2005).

At the strategic level, it is more important to represent overall system behaviour and dynamic complexity. There are fewer approaches that are readily applicable to strategic medium- to longer-term planning without (a) requiring unnecessary and unmanageable amounts of detail, most of which is not predictable at the required level, and (b) allowing service performance to be incorporated into the planning and prioritisation decisions. The system dynamics (SD) technique (Sterman 2000) lends itself naturally to situations where dynamics complexity is important, and applications of SD do tend towards the strategic end of the spectrum (Chahal and Eldabi 2008; Brailsford et al. 2010; Morecroft and Robertson 2005).

However, application of the more strategic methodologies is not without its challenges. One of these challenges may be described as 'the lure of excessive detail'. The issue of detail complexity, e.g. in DES models, is often presented mainly as a practical problem in terms of getting enough data and having enough computational power to run the models (Chahal and Eldabi 2008; Brailsford et al. 2010). However, there are fundamental as well as practical issues with modelling processes in great detail. The perils of trying to include lots of detail in an operational model are well illustrated by recent 'process mining' research that maps how processes are *actually* executed rather than how they are designed. Figure 4.1 shows a typical example (Taylor et al. 2012) where, instead of a straightforward sequential movement through the process stages, there are plenty of loops, repetitions and short-cuts. As-designed executions were only a small percentage of the overall executions of the processes analysed; all the remaining process instances deviated significantly from the original process (Taylor et al. 2012). When modelling the performance of this process, adding details could easily become self-defeating as this would require the introduction of simplifying assumptions about workflows that do not hold in reality – by adding details, the model is likely to become less not more accurate.

Another barrier to adoption of strategic modelling tools, like SD, is a lack of awareness of the importance of dynamic complexity, which is not as widely appreciated as the need to deal with the details. Operations exist in an environment that interacts with these operations, giving rise to feedback loops. Specifically, operational processes exist within an organisation which can modify them, e.g. in terms of resourcing levels, resource deployment and resourcing policies. In particular, if modelling scenarios were substantially different from the current situation, it would often be unrealistic to ignore these interactions. For example, to model how a system copes with a surge in demand above reference levels in a realistic way, there would in most cases need to be some representation of how resources would be adjusted to cope. It is possible to include such resource-adjustment feedback within DES models, (see, e.g. Wynn et al. 2012); however this seems not to be common practice.

The work described in this chapter used system dynamics to develop a high-level model of service operations. Below we show how our SD models capture dynamic complexity, and by working at an aggregate level, where we consider the overall volumes and resource usage, we aim to avoid the problems associated with

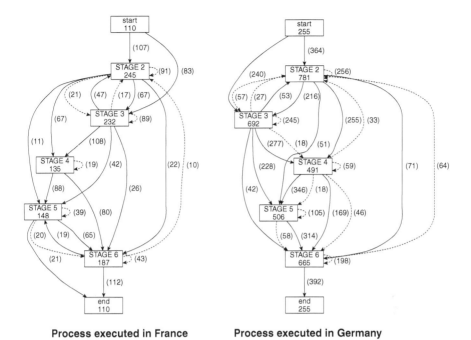

Process executed in France   **Process executed in Germany**

Fig. 4.1 Two maps of actual process execution for what is nominally the same process in two different geographies, generated by process mining (Taylor et al. 2012). The *boxes* indicate process stages with *arrows* showing movements between stages. *Dashed lines* indicate where a particular step has been taken more than once in an individual execution of the process. The numbers count instances of steps and transitions

excessive detail; the models do not rely on assumptions about the detailed paths of work through the organisation.

4.2.3   Methodology: Performance, Targets and Tension

For any service organisation, service performance from a customer viewpoint is related to both service quality and timeliness. In this chapter, we focus on cycle time—the time between customer requesting a service (installation, modification and repair) and the task being completed – as a measure to service performance. Several other performance metrics are closely related to cycle time. For example, by Little's law (see below), cycle time is related to the size of workstacks or backlogs. We have also found empirically that on-time delivery measures (fraction of tasks completed to a given deadline), in many cases, can be linked to the cycle time. Thus, even though organisations in some cases may place greater emphasis on backlogs or on time delivery measures, our focus on cycle time in how we present the methodology and results does not represent a major loss of generality.

Fig. 4.2 Workflow through
an organisation

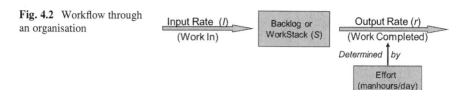

Figure 4.2 shows a simple model of workflow through an organisation. Work arrives at an average input rate I and is completed at an output rate r. At any time, some work is in progress or waiting in queues—this is the backlog (S). Little's law states that at *steady state*, the actual cycle time T_a is given by (see, e.g. Cooper 1981)

$$T_a = \frac{S}{I} = \frac{S}{r} \tag{4.1}$$

where S is the average backlog. Thus, in a steady state, cycle time is closely related to the size of the backlog.

Away from a steady state, any imbalance between input and output rates will lead to either an increase ($I > r$) or decrease ($I < r$) in the backlog, which in turn leads to an increase or decrease, respectively, in average cycle time. However, managers have no direct control over backlogs. Instead backlogs have to be managed via the output rate and the intake, with the output rate, which is determined by the level of resources deployed, normally being the main point of control.

Management of operations is thus concerned with two issues: (a) making output keep up with input and (b) reducing the mismatch between target and actual cycle times. These two priorities are separate. Much planning only focuses on (a), i.e. aiming to ensure that there is sufficient resource to deal with the expected intake. This makes the implicit assumption that backlogs remain constant; the cycle times are determined by whatever this backlog happens to be. Thus, service performance is not considered explicitly in a simple output-equals-input planning approach.

However, our hydraulics approach takes into account both of the planning considerations, which allows us to model not just the balance between inputs and outputs but also how backlogs (and hence cycle times) are affected by changes in demand and resource levels.

Figure 4.3 illustrates the approach, using system-dynamics-style diagramming. The target output rate is a weighted average of the target output rates associated with the two planning priorities discussed above:

- Output matching input: target $= I$.
- Meeting the target cycle time: target $= S/T$ where T is the target cycle time.

Thus, the target output rate, $\widetilde{r}$, is

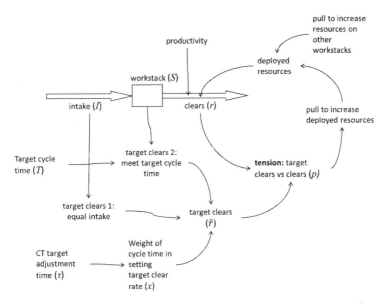

Fig. 4.3 Diagrammatic representation of the model of target setting and resource adjustment as described in the text

$$\widetilde{r} = \frac{xS}{T} + (1-x) \times I \qquad (4.2)$$

where x is the weight of the S/T term. This Eq. (4.2) can be rewritten in terms of a time constant $\tau = T/x$:

$$\widetilde{r} = \frac{S}{\tau} + \left(1 - \frac{T}{\tau}\right) \times I \qquad (4.3)$$

Under various simplifying assumptions, it is possible to show that this approach will make the actual cycle time approach the target exponentially with a time constant of τ. Thus, this time constant expresses how quickly the organisation aims to move towards the target and is an implicit expression of the importance of this target.

The comparison of actual and target output rates defines a 'tension' or a 'pull' for resources:

$$p = \frac{\widetilde{r}}{r} \qquad (4.4)$$

where r is the actual output rate and p is the 'tension' factor. This Eq. (4.4) defines the tension factor associated with reactive workstreams, e.g. repair or installation orders. Similar tension factors can be defined for nonreactive work, such as overhead activities (training, meetings, leave, etc.), and for the use of alternative resourcing options, such as overtime (where the target usually is to keep this as low as possible).

Each of the workstreams competes for resources; the key hypothesis behind our approach is that the organisation adjusts resource levels so as to equalise the 'tension' factors for each of the workstreams.

4.2.4   Testing the 'Tension' Hypothesis

We can test this hypothesis: It is possible to calculate the tension factors from actual operational data for different workstreams. If the 'tension' hypothesis is valid, then these tension factors should track each other over time. Figure 4.4 shows an example of such a chart, based on data from the operations discussed in the 'Case Study' section below. The chart uses the values for the T and τ parameters for each workstream used in the associated simulation model. The four lines correspond to two reactive workstreams, overhead activities and the use of overtime, respectively. The tension factors are calculated with the same parameters (T, τ) over the whole 5-year period, apart from a change in just two parameters at the beginning of year 3.

The chart shows that the four different tension factors are approximately equal when viewed over longer timescales. This in turn suggests that, when seen at a sufficiently aggregate level, all the detailed negotiations, resource bargaining and decision making basically amount to the organisation equalising tension. With reference to our earlier discussion, this demonstrates that it is the *dynamic complexity* that determines service performance and resource deployment. It also demonstrates that the simple equations presented above provide an adequate representation of this process.

This will never be an exact relationship; there will always be short-term deviations due to perturbations, where the system temporarily gets pushed out of balance. Also, the discrete nature of individual management decisions will add fluctuations. However, the results do indicate that the tension always tends to even out.

This fundamental insight provides a strong underpinning for the 'hydraulics' approach. The formulae for the tension factors and the assumption that the tensions move towards equality do not follow directly from the stock-flow structures or any mathematical theory. It is not something that can be derived theoretically like, for example, Little's law from queueing theory. Instead it is a hypothetical representation of *management decisions* that reflects how the 'pain' is being felt across the operations when under-resourced (or how the 'glut' is shared when over-resourced). This can never be proven exactly and can only be validated by comparing to what actually happens in real organisations.

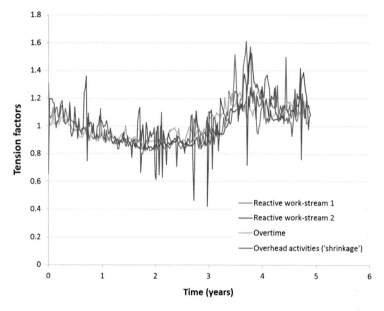

Fig. 4.4 Tension factors calculated from actual operation data

4.2.5   Modelling Resource Adjustments

The preceding sections demonstrated the tendency for tensions across an organisation to approach equality over longer timescales. This is brought about by the adjustment of resource deployment. But in order to model this resource adjustment dynamically, we need equations to describe how resources are moved between different workstreams.

In the hydraulics models, we use a first-order feedback model, familiar from the system dynamics literature (see, e.g. Sterman 2000, Chap. 8). The movement of resources between workstreams (labelled i and j) is

$$\frac{dr_i}{dt} = \sum_j \max\left(0, \frac{(p_i - p_j)r_j}{x_{ij}}\right) - \max\left(0, \frac{(p_j - p_i)r_i}{x_{ji}}\right) \tag{4.5}$$

where
  p_i is the tension factor for workstream i
  x_{ij} is the time constant for moving from i to j
  r_i is the output rate for workstream i (resources deployed to i)
  Similar equations dictate how contingency resource options, e.g. use of overtime, are deployed.

These equations introduce additional parameters into the model, viz. the time constants x_{ij}. These are the time constants for the net movement of substantial levels of resources between workstreams, not the time it takes for a single unit of resource (e.g. a field engineer) to move between one type of job and another. In some cases, we have been able to estimate the time constants through time series analysis. Discussions with managers of the organisation can also provide a guide to the values. However, a certain degree of fine-tuning is usually required in order to get good agreements between historical actuals and the models. In doing this, we try to keep the number of adjustable parameters to a minimum by assuming that many of the time constants are equal, e.g. we usually assume that $x_{ij} = x_{ji}$ for all values of i and j unless we have a good reason to suspect otherwise.

4.3 Simulation Experiments

4.3.1 Simplified Model

In this section, we use a simplified version of the full hydraulics model to demonstrate the types of scenarios that can be analysed and to illustrate the organisational behaviour encapsulated in a typical model. The simplified model focuses on overtime, the manipulation of overheads and the use of short-term contractors which tend to be the most commonly used sources of flexibility.

Figure 4.5 illustrates the structure of the model in which the field force may be deployed on one of two workstreams: provision and repair. In addition, some time will be required for the overhead or shrinkage activities as described above. Similar structures apply to overtime although not shown explicitly in the diagram.

As demand varies, the numbers of workforce (represented as full-time equivalents, or FTE) deployed between repair, provision and overhead (shrinkage) activities will change.

In all cases, the adjustment of resource levels depends not only on the comparative tensions between workstreams but also on time constants that vary with the different types of flexibility; see Eq. (4.5). Thus, for example, overtime can be changed relatively quickly (within a few days or less), whilst it may take several weeks to adjust shrinkage levels.

4.3.2 Simulation Results

4.3.2.1 Overview

This model can easily be used to simulate a large range of 'what-if' scenarios. In this section, we present some idealised scenarios that show how the organisation can respond to changes in demand or in resourcing policy. These represent typical system behaviour and illustrate how dynamic simulations supplement more

Fig. 4.5 Schematic version
of the model. FTE refers to
full-time equivalent
workforce

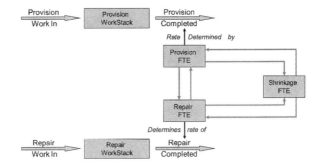

conventional planning models that simply balance average demand and capacity
(resource levels).

In the simplest case, the workflow is in a steady state where the input rate of
work (as measured in man-hours) is constant and equal to the completion rate.
Workstacks are constant throughout the simulation and correspond to the target
average cycle time [by Little's law—Eq. (4.1)].

The first two examples demonstrate how the organisation reacts to a single
perturbation to this simple case: a pulse in demand and a change in target cycle
time. Although idealised, these examples show that it can take a significant time for
the organisation to recover from a perturbation and this is a good measure of the
agility of the organisation. The third example looks at the more realistic situation
with variable demand throughout the simulation.

4.3.2.2 Spike in Demand

In this scenario, demand for repairs increases by 50 % for one week. Although
unlikely on a national scale, similar spikes might occur locally, e.g. as a result of a
severe weather event. Intake and completion rates are shown in Fig. 4.6. Completion rates increase in response to the spike in demand, but it takes several weeks to
return to the steady state.

In this initial scenario, no overtime is allowed. The only flexibilities are variation
in shrinkage (overhead activity) levels and the movement of the staff from provision to repair (Fig. 4.7). However, the ability of the organisation to respond in this
way is limited due to the competing demands from provision and shrinkage
activities.

As a result, the impact on cycle tim.es is long-lasting (Fig. 4.8). At its peak,
average cycle time is ~60 % higher than normal, and it takes several months to
return to the target cycle time. The impact on provision, due to resources being
pulled across to do repairs, is smaller but still significant. This result emphasises the
importance of modelling operations as a whole, since changes in one area inevitably
will have knock-on impacts in other areas.

A change in policy to allow some overtime (Fig. 4.9) greatly reduces the impact
of the spike and the time for the organisation to return to steady state (Fig. 4.10).

Fig. 4.6 Time variation of intake and completion rates for repairs

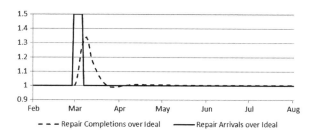

Fig. 4.7 Changes in FTE allocation in response to spike in demand

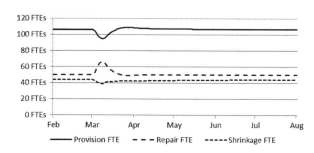

Fig. 4.8 Impact of demand spike on cycle times, represented as ratio to target (so on-target = 1)

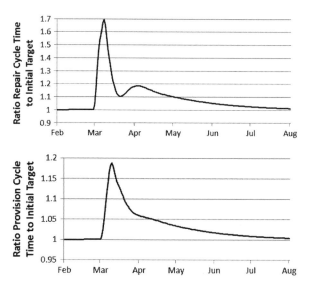

The 'over-shoot', i.e. the dips in cycle times below target, reflects the time it takes for the resource deployment to respond to the drop in intake at the end of the spike.

4.3.2.3 Reduction in Target Cycle Time

A similar response is seen if a policy decision is taken to reduce the provision cycle time (e.g. by 33 %). In this case, there is no variation in work intake. However, the

Fig. 4.9 Variation in
overtime in response to
spike in demand

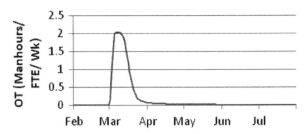

Fig. 4.10 Variation in
cycle times when overtime
is allowed. Both maximum
cycle time and time taken to
return to normal are greatly
reduced compared with
Fig. 4.8

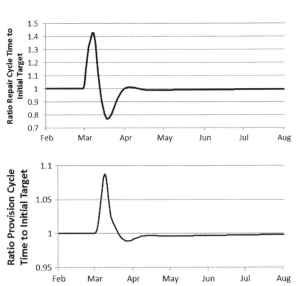

impact is similar to a spike in demand. This is because a reduction in cycle time requires a reduction in the size of the workstack. Consequently, completion rate must be increased temporarily (Fig. 4.11).

Figure 4.12 shows how repair and provision cycle times change. The initial reduction in provision cycle time is very rapid, but this slows as repair cycle times increase creating competition for resource.

Changes in FTE allocation are shown in Fig. 4.13. As before, staff moves dynamically between provision, repair and shrinkage in response to the changing demands.

Figure 4.14 shows overtime levels—these remain high for several weeks. The actual timescale depends on what limits are placed on overtime. In this example, there is a policy to limit overtime to a maximum of 3 h/FTE/week. A higher limit would shorten the time needed for the workflows to return to steady state.

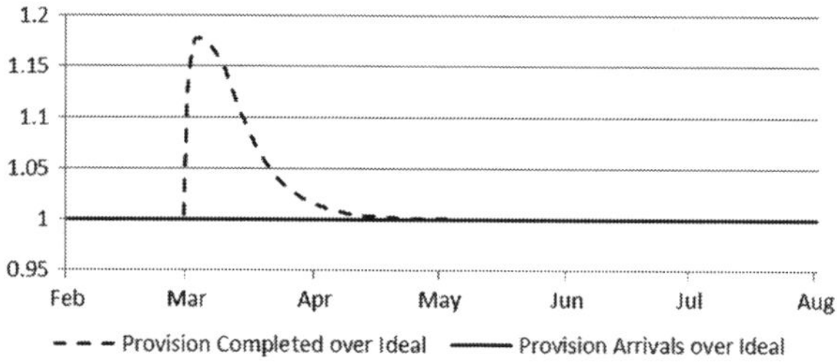

Fig. 4.11 Provision intake and completions rates for scenario where provision cycle time is reduced by 50 % from the beginning of March

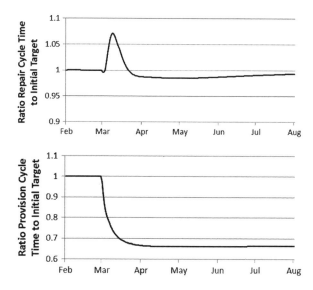

Fig. 4.12 Variation of repair (*top*) and provision (*bottom*) cycle times

Fig. 4.13 Changes in FTE allocation in response to reduction in provision cycle time

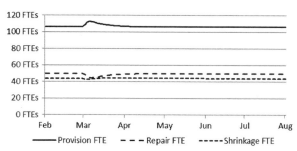

4.3.2.4 Demand Variability

The pattern of intake over time in the two examples presented above is much idealised. In reality, organisations are faced with variable demand on many timescales—weekly, daily or even hourly. A more realistic simulation is shown in Fig. 4.15, where weekly input rate varies randomly around an average value Values are shown as a ratio to the average input rate.

Fig. 4.14 Overtime used in response to reduction in provision cycle time

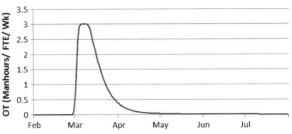

Fig. 4.15 Work in and completion rates for a more realistic simulation where input varies randomly about the average value: shown for repair (*top*) and provision (*bottom*)

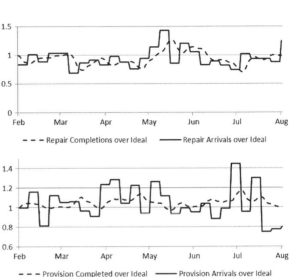

In this initial scenario, staff can transfer between the provision, repair and shrinkage workstreams but no overtime is permitted. It can be seen (Fig. 4.16) that there is considerable movement between workstreams in response to the variations in demand.

If the only concern is balancing demand and resource, then this might seem to be an acceptable outcome. Cycle times (Fig. 4.17) are close to target at both the beginning and end of the simulation, so on average there is sufficient people to meet the demand.

However, Fig. 4.17 also shows considerable variation in the cycle times; these can exceed the targets for significant periods of time. For example, in this particular run, the repair cycle time is more than 50 % higher than target throughout much of May and June. Clearly, more resource is required at times if the organisation is to meet its customer service targets. As with the examples above, an additional source of flexibility is overtime. This can greatly reduce the likelihood of cycle times exceeding target (Fig. 4.18).

Comparison with Fig. 4.17 shows that although cycle times still occasionally exceed the target (e.g. in May), the increase is much smaller and lasts for a much

Fig. 4.16 Variation in FTE numbers in response to variable demand

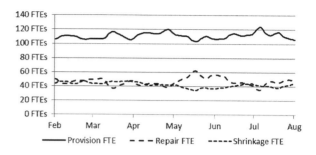

Fig. 4.17 Cycle times (as ratio to target) for repair (*top*) and provision (*bottom*)—no overtime

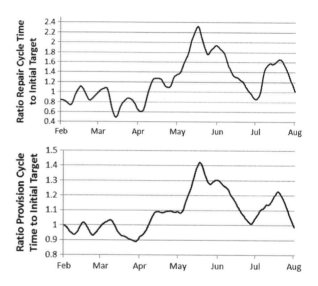

Fig. 4.18 Cycle times (as ratio to target) for repair (top) and provision (bottom)—with overtime; up to 4 h/FTE/week

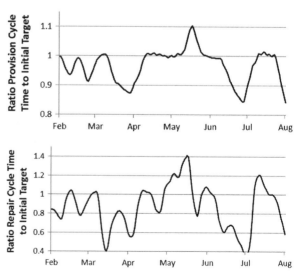

Fig. 4.19 Overtime levels in response to variations in demand

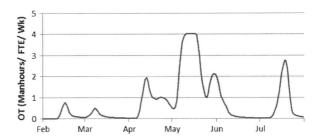

shorter period of time than when there is no overtime. However, Fig. 4.19 shows that achieving this can require sustained and significant overtime. This overtime is driven by the need to meet specific cycle times; if cycle times can be relaxed, then less overtime is required. This is a good illustration of the trade-off between costs (incurred due to overtime) and service performance (cycle time), which is driven by the dynamics of the operations, as modelled by the approach described in this chapter.

4.3.3   Uncertainty Analysis

The scenario presented in Figs. 4.15, 4.16, 4.17, 4.18 and 4.19 looked at a specific sequence of input data with week-by-week variation in intake. However, in reality these variations are stochastic. Although the results above demonstrated some of the effects of variability on performance, examining just a single scenario does not evaluate the full range of possible outcomes. A more powerful, if computationally more intensive, approach is to combine the simulation model with uncertainty analysis, running multiple scenarios through the model and subsequently aggregating the results from all the runs. The method employed here uses Latin hypercube sampling to generate multiple time series for the intake variables, as implemented in the Powersim Studio software.[1] This is similar to the more familiar Monte Carlo analysis but uses systematic selection of random variables, based on a given distribution, rather than the random sampling employed in Monte Carlo.

In the results from a multi-run uncertainty simulation shown in Fig. 4.20, we assume a policy that limits overtime to < 4h/FTE/week. Figure 4.20 shows that with this policy in place, the organisation has a ~95 % chance of meeting its service level target at any time.

But how much overtime is required? Again, this will depend on the specific demand scenario, but the Monte Carlo simulation can indicate likely requirements. Figure 4.21 shows cumulative overtime levels (expressed as hours/FTE). In 50 % of the runs, cumulative overtime at the end of the 6-month simulation is < ~10 h/FTE over a 6-month period, whilst there is a 5 % probability that it could exceed ~27

---

[1] http://www.powersim.no

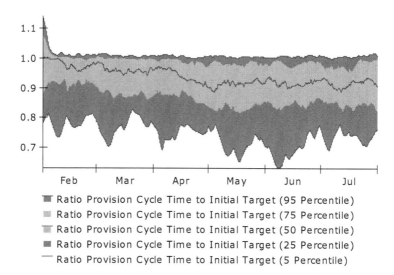

Fig. 4.20 Ratio provision cycle times to target (on target = 1). The upper limit of each percentile band corresponds to the probability (in per cent) that the ratio is below that limit at each point in time

h/FTE. Ten hours/FTE over a 6-month period may not seem a great deal, but in a team of 100, this would be enough to raise questions about whether it is better to rely on overtime or recruit another team member. There is no clear answer, but the simulations allow managers to better understand the costs of a given policy, both financially and in terms of customer experience.

4.4 Case Study: Modelling the Field Operations of a Major Line of Business

The results presented above from a fairly simple model demonstrated some of scenarios that can be explored with the hydraulics approach. In this section, we discuss the substantially more complex model developed to support planning of field operations in one of BT's major lines of business. A description of an earlier version of this model has been published previously (Jensen et al. 2006).

The model is implemented using the Powersim Studio system dynamics modelling software, but the setting of many of the inputs and the main presentation of results are done using Microsoft Excel spreadsheet files linked to the system dynamics model.

The model is a fairly detailed implementation of the 'tension' approach which includes:

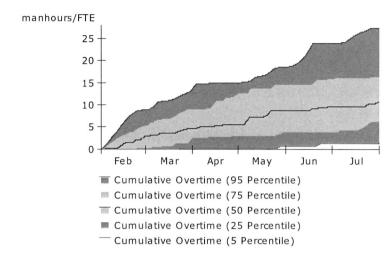

Fig. 4.21 Monte Carlo simulation—cumulative overtime

- Multiple workstreams, e.g. field repair, field installation (multiple products), exchange-based activities and capital investment activities
- Overhead activities (shrinkage)
- Overtime
- Multiple pools of resources, some single skilled, some multiskilled

Simulation is carried out at a daily level, which requires overlaying the gradual changes in parameters over time with weekly patterns, e.g.:

- Job intake (e.g. more repair jobs arrive on weekdays than at weekends)
- Scheduling of work (more on weekdays, less at weekends)
- Attendance patterns for engineers
- Scheduling of overtime (more at weekends, where fewer engineers are scheduled to work according to standard attendance patterns)

The fluctuation of variables across the weeks also necessitates care in how aggregate parameters are calculated and reported. For example, we have found that the most accurate application of Little's law to calculate average cycle times is obtained by taking 7-day rolling averages of both the workstacks and completion rates before calculating the ratio to get the estimated cycle time.

As the main inputs, the model uses time series of internally reported volumes of work, productivity parameters and resource levels (people numbers) using actuals for validation and forecasts to explore future service performance scenarios. Model parameters such as target cycle times (T), target-adjustment times (τ) and resource adjustment time constants (τ_a) are determined through a mixture of discussions with managers in the client organisation, time series analyses and adjustment to obtain agreement between the model and historical actuals.

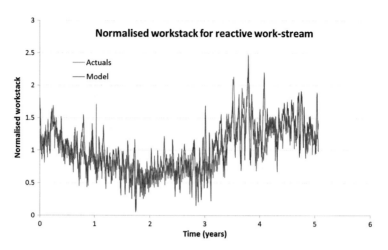

Fig. 4.22 Comparison of actual and simulated workstack levels

Figure 4.22 shows a typical result from the model, where actual workstack data is compared with output from the model. There is good agreement between model and actuals not only in terms of overall trends but also the size and shapes of peaks and dips. Being able to demonstrate that this level of agreement can be obtained across all the key performance and resourcing metrics with a small number of parameters over periods of several years is important in building confidence with clients in the accuracy of the model.

The model has been developed by our team in the research area of BT working closely at various points both with senior managers and with business planners and analysts within the client organisation. It has been used over several years undergoing many changes along the way, both to improve functionality and usability and to reflect organisational changes.

Typical applications have included:

- Testing resourcing scenarios to give costs vs. service performance estimates, which no other model was available to do. Results were used by senior managers in strategic planning dialogues.
- Analysing ad hoc scenarios, such as surges in demand or temporary resource shortages, looking to evaluate how badly service performance might be affected and how quickly the system would recover.
- Analysing service recovery dependent on prioritisation, including estimates of the knock-on effects on other parts of organisation if one workstream is prioritised to get it back on target.
- Risk analysis—similar to the approach described above. The analyses looked at both forecast uncertainty (uncertain of overall levels of demand) and fluctuations in intake.

4.5   Challenges and Limitations

Although our hydraulics modelling has been a success—the longevity of the project is a testament to that— it is only fair to acknowledge that there have been challenges as well.

Our approach, in common with a lot of other SD modelling, relies on a fair degree of abstraction. For example, the 'tension' concept is less tangible than the objects you would see in a DES model. Other authors (Brailsford and Hilton 2001), when doing a comparison with DES, have observed that quantitative SD models with their stock-flow diagrams and accompanying differential equations are less likely to be understood by clients. This means that there is always a period of familiarisation with the methodology required in a modelling project and—most importantly—the (potential) clients have to be comfortable with using an unfamiliar, and to some baffling, approach.

There is also a constant tension between the amount of detail that is feasible and desirable to include in models and expectations of clients. With reference to the discussion above about dynamic complexity and detail complexity, because detail complexity is often perceived as the main barrier to getting to grips with a problem, requests for modelling work often come packaged with requirements for a lot of details. In some case, this may of course be a legitimate need, but often you get the demand for detail even when—from our perspective—more accurate and meaningful answers could be obtained without this detail. Again, a certain amount of client education and negotiation is usually required to overcome this issue, but in some cases, we just had to acknowledge that the client was too set on having lots of details and that our approach would never be right for this particular client.

The use of a pure SD approach does carry technical limitations. For example, we can model backlogs and from this calculate average cycle times, but we cannot model the distributions of cycle times across the population of tasks. Thus, we cannot directly model the fraction of tasks that would fail deadlines, which in some cases is more important to the organisation than the averages. As explained previously, in some cases we can get around this by employing empirically derived relationships between the averages and the threshold measures, but not always, and it would be desirable to able to model these quantities directly. A combined DES and SD approach may be able to overcome this shortcoming. Others (Chahal and Eldabi (2008) and Brailsford et al. (2010) have attempted to develop such a methodology, but no well-established, mature approach exists. Thus, it will require future research on our part to develop such a methodology and to find an implementation that will work for the type of problems described, i.e. the modelling of performance in service operations.

4.6   Conclusions

In this chapter we have described an approach to modelling service operations appropriate to long-term strategic planning which incorporates the modelling of service performance as an integral feature of the methodology. In positioning the approach, we stressed the importance of emphasising *dynamic* complexity over *detail* complexity. We specifically argued that it is necessary to include controlling feedback loops for resource deployment and also to avoid excessive detail, which may not only be cumbersome but also misleading.

We have used the system dynamics (SD) technique as a starting point for this so-called 'hydraulics' approach. At the core of the hydraulics approach is the observation that the combined effect of decisions, e.g. on resource deployment, works to equalise tension across the organisation with 'tension' represented by equations relating performance, targets and prioritisations.

Using a fairly simple implementation of this approach, we demonstrated typical behaviours of hydraulics models and showed some of the alternative scenarios that can be explored. This was followed by a description of how we have use a more comprehensive version, validated against historical data, to model the field operations of a major BT line of business describing how the model has been used and also some of challenges faced.

We are constantly refining and improving this methodology, e.g. seeking to incorporate useful aspects of more detailed modelling methodologies and to enable greater integration with data-driven statistical approaches, and thus generally aiming to bridge the gap between strategic modelling, as described in this chapter, and more tactical planning approaches, as described elsewhere in this book.

References

Brailsford S, Hilton N (2001) A comparison of discrete event simulation and system dynamics for modelling health-care systems. In: Riley J (ed) Proceedings of the 26th meeting of the ORAHS Working Group 2000. Glasgow Caledonian University, Glasgow Scotland, pp 18–39

Brailsford SC, Desai SM, Viana J (2010) Towards the holy grail: combining system dynamics and discrete-event simulation in healthcare. In: Johansson B et al (ed) Proceedings of the 2010 winter simulation conference, pp 2293–2303

Chahal K, Eldabi T (2008) Applicability of hybrid simulation to different modes of governance in UK healthcare. In: Mason SJ, et al (ed) Proceedings of the 2008 winter simulation conference, pp 1469–1477

Cooper RB (1981) Introduction to Queueing theory, 2nd edn. New York, NY, 2006

Dietrich B (2006) Resource planning for business services. Commun ACM 49:62–64

Forrester JW (1960) The impact of feedback control concepts on the management sciences. In: Collected papers of J.W. Forrester. Wright-Allen Press, Cambridge MA, 1975, pp 45–60

Jensen K, Barnsley P, Tortolero J, Baxter N (2006) Dynamic modelling of service delivery. BT Technol J 24:48–59

Morecroft J, Robertson S (2005) Explaining puzzling dynamics: comparing the use of system dynamics and discrete-event simulation. In: Conference proceedings, the 23rd International Conference of the System Dynamics Society, Boston, 17–21 July 2005

Oliva R, Sterman JD (2001) Cutting corners and working overtime: quality erosion in the service industry. Manag Sci 47:894–914

Oliva R, Sterman JD (2010) Death spirals and virtuous cycles: human resource dynamics in knowledge-based services. In: Maglio P, Spohrer J, Kieliszewski C (eds) The handbook of service science. Springer, London, pp 321–358

Sasser WE Jr (1976) Match supply and demand in service industries. Harvard Bus Rev 54:133–140

Senge P (1990) The fifth discipline: the art and practice of the learning organization. Century Business, Irvine, CA, pp 71–72

Sterman J (2000) Business dynamics: systems thinking and modeling for a complex world. Irwin/ McGraw-Hill, Boston, MA

Taylor P, Leida M, Majeed BA (2012) Case study in process mining in a multinational enterprise. In: Lecture notes in business information processing 0116, pp 134–153

Wynn DC, Cassidy S, Clarkson, PJ (2012) Design of robust service operations using cybernetic principles and simulation. Paper presented at International design conference – Design 2012. Dubrovnik, Croatia. May 21st–24th 2012

Chapter 5
Understanding the Risks of Forecasting

Jonathan Malpass

Abstract A fundamental element of the successful deployment of operational teams is the ability to forecast future demand accurately. There is a raft of forecasting methods available to the forecaster ranging from the simple to the highly sophisticated, with many software packages able to identify the most appropriate method. One of the skills of the forecaster is the ability to strike a balance between finding a model that is good enough and one that overfits the data, especially if there is a demand from the business to seek forecasts with greater accuracy. However, the pursuit of better forecasts can sometimes be a fruitless exercise and Is it important for any forecaster to ask two questions: firstly, Is it of benefit to improve accuracy, and secondly, given the data available, can better forecasts be produced? The first question is a matter of identifying the effort required to produce better forecasts, be that through identifying a better model or collecting additional data to improve the existing model, and weighing it against the improvements in accuracy. The second question is a matter of understanding the limitations of the current forecasting model and, as importantly, the data used to produce the forecasts. If, for instance, the data contains a high level of randomness, even the most sophisticated model will be unable to produce highly accurate forecasts. This chapter describes an approach based on forecast errors to derive an approximation of the signal-to-noise ratio that can be used to understand the limitations of a given forecasting method on a given set of data and therefore provides the forecaster with the knowledge of whether there is any likely benefit from seeking further improvements. If the forecaster accepts the current model, the signal-to-noise ratio can provide an understanding of the risks associated with that model.

J. Malpass (✉)
Research and Innovation, BT Technology, Services and Operations, Martlesham Heath, UK

G. Owusu et al. (eds.), *Transforming Field and Service Operations*,
DOI 10.1007/978-3-642-44970-3_5, © Springer-Verlag Berlin Heidelberg 2013

5.1 Introduction

The need to forecast accurately is imperative for any service organisation; the associated costs of deploying too many resources or the impact of failing to meet service levels means that poor forecasts have a direct impact on an organisation's bottom line. Consequently, many organisations invest a great deal of time and money into ensuring that their forecasts are as accurate as possible, and there is a wide choice of forecasting approaches that can be employed in an attempt to forecast levels of demand in order to deploy sufficient resources to meet service levels (Malpass and Shah 2008).

A common misconception, however, is that forecasts are inaccurate because the statistical method, or the general methodology, is inappropriate or being applied incorrectly. A question regularly heard by this author is the following: "Why aren't our forecasts more accurate?"

The answer to this question often has nothing to do with the method or the underlying model, but more to do with the data being gathered. This is not to suggest that the quality of the data is poor. On the contrary, the data is often highly reflective of the situation. Inaccurate forecasts are, quite often, solely to do with the variability within the data.

All measurable phenomena are subject to some degree of variation, be it the weight of a loaf of bread, the number of hours that a light bulb lasts or the time taken to perform a task. Statistical process control (SPC) (Oakland 1996) can be used to identify "special causes" of variation, i.e. those events that have some attributable cause which can explain why particular observations and patterns, such as the rise or fall over time, occur. Similarly, SPC can be used to understand the "common cause" variation present in the data, i.e. "noise". This variation is, by definition, impossible to model.

Forecasters who try to forecast "noise" will inevitably fail and spend an inappropriate amount of time trying to increase accuracy. By understanding the level of noise in the data, the forecaster can manage the risks that will be associated with imperfect forecasts and a manager can build some tolerance, or contingency, into a process. Many forecasters seek to apply prediction intervals to their forecasts, but not all methods allow for this. Understanding the noise in the data allows for a range of forecast errors to be estimated and therefore the risk can be mitigated.

5.2 Understanding a Time Series

A time series (A) is a sequence of data points representing a phenomenon of interest measured at equally spaced, uniform time intervals. It can be said to comprise two elements: pattern (or signal), which can be modelled or explained by one or more factors, and error (or noise) which is random and cannot be modelled, i.e.

$$A = f(pattern, error) \text{ or } A = f(signal, noise) \qquad (5.1)$$

Typical patterns include trend, seasonality and cyclical events, thus

$$A = f(trend, seasonality, cycle, error) \qquad (5.2)$$

A relatively simple approach to forecasting is to decompose the time series by modelling each of the separate factors. Performing this decomposition enables the forecaster to remove known patterns from the data, thus leaving the random error. Good statistical and data mining packages, such as Minitab (2010), SPSS (2010) and KXEN (2012), have procedures that automatically decompose time series into component parts.

Figure 5.1 shows an example of how a time series can be decomposed: the data exhibits both trend and seasonality (Fig. 5.1a). Once the trend is identified (Fig. 5.1b), a more complete picture of the seasonality is obtained (Fig. 5.1c). On removing this factor, the remaining variation is the error (Fig. 5.1d).

5.2.1   Forecasting Methods

This chapter is not concerned with identifying the best forecasting technique for a given data set. There are many such methods, including decomposition, exponential smoothing methods, regression, ARIMA and many others which are explained in great detail elsewhere (Makridakis et al. 1998).

The concept central to this chapter is that the signal in the time series can be modelled and the noise that remains can be quantified. Thus, the decomposition of a time series into pattern and error is a useful exemplar.

5.2.2   Measuring Forecast Accuracy

There are many measures of forecast accuracy, and as with particular forecasting methods, this chapter is not concerned with rating one measure over another. Excellent reviews (Armstrong and Collopy 1992; Hyndman and Koehler 2006) have already shown that measures such as the mean absolute percentage error (MAPE) and mean squared error (MSE) have their merits. The similarity percentage (SP) (Bunn and Taylor 2001) is of greatest utility for the purposes of this work.

If the series of data being forecast is A and the forecast is F, then the similarity percentage for the forecast at time t is defined as

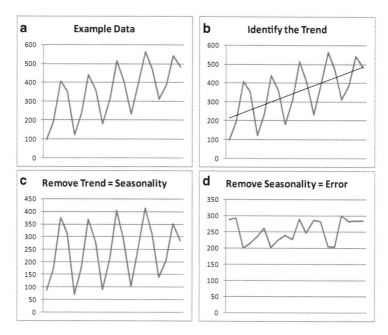

Fig. 5.1 Decomposition of a time series into trend, seasonality and error

$$If\ A_t < F_t, SP_t = \frac{A_t}{F_t}; If\ A_t > F_t, SP_t = \frac{F_t}{A_t} \tag{5.3}$$

$$SP = \sum_{t=1}^{n} SP_t \tag{5.4}$$

The *SP* approaches 1 (i.e. 100 %) as greater forecast accuracy is achieved. To aid interpretation and to enable easy comparisons with measures such as MAPE, the level of error is found to be more convenient, i.e. $1 - SP$

5.3 The Signal-to-Noise Ratio

5.3.1 A Definition

The signal-to-noise ratio (SNR) is a measure used in science and engineering to quantify how much the desired signal has been corrupted and can be defined as the ratio of the level of a desired signal to the level of background noise. While SNR is commonly quoted for electrical and audio signals, it can be applied to any form of signal, including any phenomenon in which a pattern exists such as process cycle time or sales figures.

A generic definition of the signal-to-noise ratio, SNR_g, (Signal To Noise Ratio 2008) is the reciprocal of the coefficient of variation (CoV), i.e.

$$SNR_g = \frac{1}{CoV} = \frac{\mu}{\sigma} \qquad (5.5)$$

where μ is the mean and σ is the standard deviation of the time series A. A value greater than 1 indicates that there is more signal than noise and the higher the ratio, the less obtrusive the background noise is.

Using this definition, the signal-to-noise ratio increases as the standard deviation decreases and tends to infinity as variation diminishes to zero. For convenience, the definition used here is that the SNR is the coefficient of variation, i.e.

$$SNR = CoV = \frac{\sigma}{\mu} \qquad (5.6)$$

Thus, smaller values of SNR indicate lower levels of noise in the process, and the SNR tends to zero as the standard deviation decreases.

The concept of the SNR in forecasting is not novel. Simple linear regression can be used for forecasting a time series, and it has been suggested that information about the precision of the regression estimates can be interpreted as a signal-to-noise ratio (Decision 441 Forecasting 2005).

5.3.2   A Simple Example

Consider two office workers who monitor the time it takes for them to commute to work every day for four weeks (see Fig. 5.2). The first commuter finds that their journey time is, on average, 30 min with a standard deviation of 3.5 min. The signal-to-noise ratio [Eq. (5.6)] is thus $3.5/30 = 0.118$. A colleague finds that their average commute is 40 min, with a standard deviation of 6.5 min. The signal-to-noise ratio is, therefore, $6.5/40 = 0.163$.

In both cases, there is no discernible pattern to the data and thus forecasting the commuting time for a given day is subject to a random element. Using the average of each person's daily commute as a forecast yields a reasonable level of accuracy: $1 - SP$ for commuter 1 is 9.6 % and 13.7 % for commuter 2. Clearly commuter 2 has greater error in their forecasts. If they only allowed 40 min to travel to work, they will frequently arrive late at the office. It would be reasonable to assume that the level noise in the data is correlated with the accuracy of the forecasts; thus the forecasts cannot be any more successful.

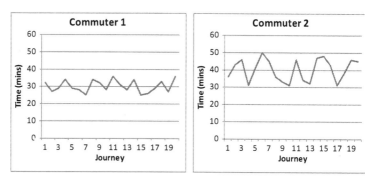

Fig. 5.2 Example of commuting times for two office workers

5.4 Understanding the Limits to Forecast Accuracy?

The aim of this chapter is to illustrate that the accuracy of forecasts is limited by the amount of noise in a time series and that for a given signal-to-noise ratio, it is possible to estimate the accuracy of the forecasts; thus a baseline level of accuracy can be provided which can be used to assess forecasting models and to enable some level of risk to be associated with future forecasts. To ensure that any baseline is reliable, it is essential that two assumptions about the SNR are true, i.e.:

- Assumption 1: the signal-to-noise ratio is correlated to forecast accuracy, $1 - SP$.
- Assumption 2: the forecast accuracy and signal-to-noise ratio are independent of the mean of the data.

5.4.1 Experimental Design

To validate these assumptions, a series of experiments were performed in which different signal-to-noise ratios were generated for data sets consisting of various numbers of observations, n. Data sets of normally distributed random numbers of different means and standard deviations were generated using the random number generation function within Microsoft Excel. Table 5.1 shows the 95 combinations used; 250 different data sets were generated for each combination. Thus, 23,750 individual data sets were generated for each value of n.

For each data set, A, the forecast for each point, A_i, is taken to be the average of that data set, i.e.

$$F_i = \overline{A} = \frac{1}{n} \sum_{1}^{n} A_i \tag{5.7}$$

Table 5.1 Combinations of means and standard deviations used in the experiment

		standard deviation																		
		6	12	17	23	29	35	40	46	52	58	63	69	75	81	87	92	98	104	110	116
mean	50	•	•	•	•	•															
	100	•	•	•	•	•	•	•	•	•	•										
	200	•	•	•	•	•	•	•	•	•	•	•	•	•	•	•	•	•	•	•	•
	300	•	•	•	•	•	•	•	•	•	•	•	•	•	•	•	•	•	•	•	•
	500	•	•	•	•	•	•	•	•	•	•	•	•	•	•	•	•	•	•	•	•
	1000	•	•	•	•	•	•	•	•	•	•	•	•	•	•	•	•	•	•	•	•

5.4.2 Limitation

The data used in these experiments are all stationary time series, i.e. the mean is constant across time. Clearly this would appear to limit the usage of the signal-to-noise ratio as many data sets contain patterns, such as trend and seasonality. However, if these patterns can be modelled and removed (e.g. seasonal data are de-seasonalised), the hypothesis can be applied to any data set.

5.4.3 Results

The first assumption is that the signal-to-noise ratio is correlated to forecast accuracy. The forecast accuracy, $(1 - SP)$, for each size of data set, n, and combination of mean and standard deviation was found. Table 5.2 shows a sample of results for the 30 observation data sets: the "Mean" and "StDev" are the average and standard deviation of all 250 averages for each combination, the signal-to-noise ratio, "SNR", is the average of all 250 data sets and the "Ave (1-SP)" is the average forecast accuracy of all 250 data sets. The "Min(1-SP)" and "Max(1-SP)" are the best and worst forecast accuracies of each combination.

For example, the data sets generated to have mean 100 and standard deviation 35 has an average SNR of 0.35, and the 250 data sets of this combination have produced forecast accuracies between 15.3 % and 34.8 % with an average of 23.6 %. Compare this to the norm (100, 12) data sets where the SNR is 0.12 and $1 - SP$ is 8.7 %, with a range of 5.6–13.1 %. This result means that not only is the "likely" accuracy for data with less noise to be better but also the range of possible accuracies is smaller. Therefore, there is less risk associated with forecasting data sets with less noise.

This pattern of results can be seen throughout the different SNR values. Figure 5.3 shows a scatter plot of the SNR against $1 - SP$ for each of the 23,750 data points from the 30 observation experiments, and Fig 5.3b shows a plot summarising the average SNR and $1 - SP$ of each of the 95 combinations of mean and standard deviation.

Table 5.2 Sample of summary statistics of the 30 observation data sets with mean 100

Distribution	Mean	StDev	SNR	Min (1-SP)	Ave (1-SP)	Max (1-SP)	Count
Norm(100,6)	100.0	5.8	0.06	2.4%	4.5%	6.2%	250
Norm(100,12)	100.2	11.8	0.12	5.6%	8.7%	13.1%	250
Norm(100,17)	100.1	16.7	0.17	7.2%	12.1%	16.6%	250
Norm(100,23)	100.0	22.7	0.23	11.3%	16.1%	22.7%	250
Norm(100,29)	99.5	28.7	0.29	12.9%	20.0%	26.7%	250
Norm(100,35)	100.5	34.6	0.35	15.3%	23.6%	34.8%	250
Norm(100,40)	99.7	39.7	0.40	17.8%	26.7%	38.6%	250
Norm(100,46)	99.5	45.6	0.46	18.7%	30.3%	49.6%	250
Norm(100,52)	99.8	51.7	0.52	22.0%	33.8%	54.9%	250
Norm(100,58)	100.5	57.6	0.58	21.6%	37.0%	53.4%	250

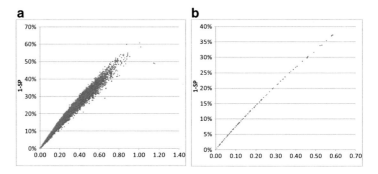

Fig. 5.3 Plot of signal-to-noise ratio against forecast accuracy (1-SP) for each of the 23,750 data sets in the 30 observation experiments (**a**) and for the summary of the 95 combinations (**b**)

The second assumption requiring validation is that the forecast accuracy and signal-to-noise ratio is independent of the mean of the data, i.e. a normally distributed data set with mean of 100 and standard deviation of 6, i.e. N(100,6), should be expected to have the same SNR as a normally distributed data set with mean of 1,000 and standard deviation of 60, i.e. N(1000,60). If Assumption 1 is generally held to be true, then the forecast accuracy of both data sets will, therefore, be the same.

Table 5.3 shows the summary of results for the experiments for data sets comprising 30 observations that yield an SNR of approximately 0.12 and 0.23. The Ave (1-SP) for the first SNR is approximately 8.6–8.8 %. The second set has an accuracy of 16.0–16.3 %. Thus it appears that the SNR and forecast accuracy are both independent of the relative size of the data.

Table 5.3 Sample of summary statistics of the 30 observation data sets with SNR values of 0.12 and 0.23

Distribution	Mean	StDev	SNR	Min (1-SP)	Ave (1-SP)	Max (1-SP)	Count
Norm(200,23)	200.5	22.5	0.11	5.5%	8.4%	13.3%	250
Norm(50,6)	50.0	6.0	0.12	5.8%	8.8%	12.9%	250
Norm(100,12)	100.2	11.8	0.12	5.6%	8.7%	13.1%	250
Norm(300,35)	299.8	34.7	0.12	5.2%	8.6%	12.5%	250
Norm(500,58)	501.0	57.8	0.12	5.4%	8.6%	12.0%	250
Norm(1000,116)	999.4	115.1	0.12	5.4%	8.6%	12.1%	250
Norm(300,69)	301.2	68.1	0.23	10.6%	16.2%	24.1%	250
Norm(200,46)	200.1	45.3	0.23	11.0%	16.0%	24.0%	250
Norm(100,23)	100.0	22.7	0.23	11.3%	16.1%	22.7%	250
Norm(500,116)	500.6	115.2	0.23	11.6%	16.3%	23.1%	250
Norm(50,12)	49.9	11.9	0.24	10.8%	16.9%	23.5%	250

5.4.4 Deriving Forecast Limits for a given Signal-to-Noise Ratio

The results in Sect. 5.4.3 demonstrate that the forecast accuracy has a very strong relationship with the signal-to-noise ratio and that both accuracy and SNR are scale independent. By combining all data sets of equal SNR (to 2 decimal places), a range of forecast accuracy can be estimated. Table 5.4 shows the average accuracy of all the data sets comprising 30 observations with signal-to-noise ratios from 0.0 to 0.10. The lower and upper values are the 2.5th and 97.5th percentiles of each set of results.

The results indicate that for a data set where the SNR is 0.07, e.g. N(100,7), the expected accuracy is about 5.3 % but could reasonably be expected to be anywhere in the range 4.6–6.0 %. This means that at any one point in time, a manager in charge of a process with data following such a distribution can expect forecasts to have an average error between 4.3 % and 6.0 % and so needs to plan for such variation. If this planning is deemed unsustainable, the only recourse is to seek improvements that remove noise from the process and thus improve forecasts.

By expanding this table, an estimate of expected forecast accuracy for any given signal-to-noise ratio for data sets comprising 10, 30, 50 and 70 observations can be made, and the results are shown in Table 5.5. For example, a data set comprising 50 observations with a SNR of 0.25 can be expected to yield a forecast accuracy of about 17.6 %, with a range of 16.1–19.1 %.

It is interesting to note that the forecast accuracy is roughly the same for any given SNR irrespective of the number of observations, with only the range of the lower and upper bounds decreasing as the data set size increases.

Table 5.4 Summary of forecast accuracy for 30 observation data sets by SNR values

S/N	30 Obs		
	Lower	Average	Upper
0.00	0.3%	0.4%	0.4%
0.01	0.4%	0.8%	1.2%
0.02	1.1%	1.6%	2.0%
0.03	1.9%	2.3%	2.8%
0.04	2.6%	3.1%	3.6%
0.05	3.3%	3.9%	4.4%
0.06	4.0%	4.6%	5.2%
0.07	4.6%	5.3%	6.0%
0.08	5.3%	6.0%	6.8%
0.09	5.9%	6.8%	7.6%
0.10	6.4%	7.5%	8.3%

5.5 What Is the Benefit to Your Business?

Understanding the level of noise in a time series will not improve forecast accuracy but it does allow for error in forecasts to be managed. The scenario described in this section illustrates how a management team can allocate sufficient resources to tolerate some of the noise in the system.

5.5.1 Example: Work Volume Forecasting

Within a resource management environment, understanding the likely volumes of work on a given day is essential for deciding how much of a resource is required in order to meet business targets. Figure 5.4 shows the number of jobs arriving at a regional work allocation team every weekday (i.e. Monday to Friday) for three months. The team has to forecast as accurately as possible so that they can ensure sufficient people are available to complete the jobs that arrive. They know that job arrivals exhibit a seasonal pattern with each day of the week showing different characteristics. From the training data (Fig. 5.4), the team can also see that the volumes exhibit a downward trend and their forecasting model takes into account both the trend and seasonality.

Despite pressure to improve the forecasts, the team is unable to find a model that sufficiently reduces error. Understanding of the signal-to-noise ratio would provide the team with the limits to the forecast accuracy.

Removing the trend and seasonality from the data and then calculating standard deviation of the transformed data give an estimate of the noise in the data. The SNR is calculated by dividing this standard deviation by the mean of the original data. This yields an SNR of 0.11 which, according to Table 5.5, corresponds to an expected error level of 8.2 %, with a potential range 7.5–8.9 % (based on 70 observations).

Table 5.5 Forecast accuracy for 10, 15, 20, 30, 50 and 70 Observation data sets by SNR values

SNR	10 obs			15 obs			20 obs			30 obs			50 obs			70 obs		
	Lower	Exp.	Upper	Lower	Exp.	Upper	Lower	Exp.	Upper	Lower	Exp.	Upper	Lower	Exp.	Upper	Lower	Exp.	Upper
0.00	0.2%	0.3%	0.4%	0.3%	0.3%	0.4%	0.3%	0.4%	0.4%	0.3%	0.4%	0.4%	0.3%	0.4%	0.4%	0.4%	0.4%	0.4%
0.01	0.4%	0.8%	1.2%	0.4%	0.8%	1.2%	0.4%	0.8%	1.2%	0.4%	0.8%	1.2%	0.4%	0.8%	1.1%	0.4%	0.8%	1.1%
0.02	1.1%	1.5%	2.1%	1.1%	1.6%	2.0%	1.2%	1.6%	2.0%	1.1%	1.6%	2.0%	1.2%	1.6%	2.0%	1.2%	1.6%	2.0%
0.03	1.6%	2.3%	2.9%	1.8%	2.3%	2.9%	1.8%	2.3%	2.8%	1.9%	2.3%	2.8%	1.9%	2.4%	2.8%	1.9%	2.4%	2.8%
0.04	2.3%	3.0%	3.8%	2.2%	3.0%	3.7%	2.4%	3.1%	3.7%	2.6%	3.1%	3.6%	2.6%	3.1%	3.6%	2.7%	3.1%	3.5%
0.05	2.7%	3.8%	4.6%	3.0%	3.8%	4.5%	3.2%	3.8%	4.5%	3.3%	3.9%	4.4%	3.4%	3.9%	4.4%	3.4%	3.9%	4.3%
0.06	3.4%	4.5%	5.4%	3.6%	4.5%	5.3%	3.8%	4.6%	5.3%	4.0%	4.6%	5.2%	4.0%	4.6%	5.1%	4.1%	4.6%	5.1%
0.07	3.7%	5.2%	6.3%	4.3%	5.3%	6.1%	4.4%	5.3%	6.0%	4.6%	5.3%	6.0%	4.8%	5.3%	5.9%	4.8%	5.3%	5.9%
0.08	4.5%	5.9%	7.1%	4.9%	6.0%	7.0%	5.0%	6.0%	6.9%	5.3%	6.0%	6.8%	5.5%	6.1%	6.8%	5.6%	6.1%	6.7%
0.09	4.9%	6.7%	8.0%	5.5%	6.7%	7.8%	5.7%	6.7%	7.7%	5.9%	6.8%	7.6%	6.1%	6.8%	7.5%	6.2%	6.8%	7.4%
0.10	5.5%	7.3%	8.7%	6.0%	7.4%	8.5%	6.3%	7.5%	8.5%	6.4%	7.5%	8.3%	6.8%	7.5%	8.2%	6.9%	7.5%	8.1%
0.11	6.2%	8.1%	9.7%	6.6%	8.2%	9.3%	6.9%	8.2%	9.2%	7.2%	8.2%	9.1%	7.4%	8.2%	9.0%	7.5%	8.2%	8.9%
0.12	6.4%	8.7%	10.2%	7.2%	8.8%	10.2%	7.5%	8.9%	10.0%	7.6%	8.9%	9.9%	8.1%	8.9%	9.7%	8.2%	9.0%	9.7%
0.13	7.0%	9.4%	11.2%	7.9%	9.5%	11.0%	8.0%	9.5%	10.8%	8.4%	9.6%	10.6%	8.7%	9.6%	10.5%	8.8%	9.6%	10.4%
0.14	7.6%	10.1%	12.2%	8.4%	10.2%	11.8%	8.8%	10.3%	11.6%	9.0%	10.3%	11.4%	9.4%	10.3%	11.2%	9.5%	10.3%	11.1%
0.15	8.2%	10.8%	12.9%	8.9%	11.0%	12.5%	9.2%	10.9%	12.3%	9.6%	10.9%	12.1%	10.0%	11.0%	11.9%	10.2%	11.0%	11.8%
0.16	8.7%	11.5%	13.6%	9.8%	11.6%	13.3%	9.9%	11.6%	13.0%	10.0%	11.1%	12.1%	10.6%	11.7%	12.7%	10.8%	11.7%	12.6%
0.17	9.2%	12.1%	14.4%	9.9%	12.2%	14.0%	10.6%	12.3%	13.8%	10.4%	12.0%	13.4%	11.1%	12.3%	13.4%	11.5%	12.4%	13.3%
0.18	9.7%	12.8%	15.2%	10.5%	12.9%	14.7%	10.9%	13.0%	14.6%	11.3%	13.0%	14.4%	11.8%	13.1%	14.2%	12.0%	13.0%	14.0%
0.19	10.3%	13.6%	16.0%	11.0%	13.6%	15.5%	11.2%	13.6%	15.3%	11.9%	13.6%	15.1%	12.3%	13.7%	14.9%	12.6%	13.7%	14.7%
0.20	10.3%	14.2%	16.9%	11.1%	14.1%	16.3%	12.0%	14.3%	16.3%	12.5%	14.3%	15.8%	13.0%	14.4%	15.6%	13.2%	14.4%	15.4%
0.21	10.7%	14.7%	17.4%	12.0%	15.0%	17.0%	12.5%	14.9%	16.7%	13.0%	15.0%	16.5%	13.4%	15.0%	16.3%	13.9%	15.1%	16.1%
0.22	11.4%	15.4%	18.2%	12.6%	15.5%	17.3%	13.1%	15.6%	17.6%	13.8%	15.7%	17.3%	14.2%	15.6%	17.0%	14.5%	15.7%	16.7%
0.23	12.0%	16.1%	19.1%	13.3%	16.2%	18.6%	13.9%	16.2%	18.2%	14.3%	16.3%	17.9%	14.8%	16.3%	17.6%	15.1%	16.3%	17.4%
0.24	12.5%	16.6%	19.8%	13.7%	16.8%	19.4%	14.5%	16.9%	19.1%	14.9%	16.9%	18.7%	15.4%	17.0%	18.4%	15.7%	17.0%	18.2%
0.25	12.9%	17.4%	20.3%	14.1%	17.4%	20.1%	14.8%	17.6%	19.7%	15.6%	17.6%	19.4%	16.1%	17.6%	19.1%	16.5%	17.7%	18.8%
0.26	13.4%	18.0%	21.3%	14.7%	18.1%	20.7%	14.9%	18.2%	20.6%	16.2%	18.3%	20.0%	16.5%	18.2%	19.7%	16.8%	18.3%	19.7%
0.27	14.1%	18.5%	22.1%	14.8%	18.7%	21.5%	16.0%	18.9%	21.3%	16.6%	18.9%	20.7%	16.8%	18.8%	20.2%	17.5%	18.8%	20.2%
0.28	14.6%	19.1%	22.7%	16.0%	19.5%	22.5%	16.6%	19.5%	21.9%	17.2%	19.4%	21.3%	17.9%	19.5%	21.1%	18.2%	19.5%	20.8%
0.29	15.2%	19.8%	23.3%	16.5%	20.1%	23.0%	16.3%	19.9%	22.6%	17.4%	20.0%	22.0%	18.3%	20.2%	21.8%	18.6%	20.1%	21.6%
0.30	15.4%	20.5%	24.2%	17.0%	20.8%	23.3%	17.8%	20.7%	23.4%	18.4%	20.7%	22.9%	18.8%	20.8%	22.6%	19.2%	20.7%	22.1%
0.31	16.2%	21.0%	25.3%	16.7%	21.2%	24.4%	17.6%	21.3%	24.1%	18.9%	21.3%	23.6%	19.2%	21.3%	23.0%	19.5%	21.3%	22.7%
0.32	16.0%	21.6%	25.7%	18.5%	21.9%	24.8%	18.5%	22.0%	25.0%	19.7%	22.1%	24.3%	20.2%	22.0%	24.1%	20.4%	22.0%	23.5%
0.33	15.9%	22.1%	26.4%	18.5%	22.2%	25.2%	19.2%	22.4%	25.3%	19.9%	22.6%	25.1%	20.5%	22.6%	24.3%	21.0%	22.7%	24.2%
0.34	15.7%	22.8%	26.7%	18.6%	23.0%	26.6%	19.8%	23.2%	26.2%	20.4%	23.3%	25.7%	21.1%	23.2%	25.1%	21.6%	23.4%	24.8%
0.35	17.8%	23.7%	27.9%	19.1%	24.0%	27.2%	20.6%	23.7%	26.6%	19.9%	22.7%	25.3%	21.7%	23.8%	25.9%	22.0%	23.8%	25.6%
0.36	17.4%	24.2%	28.6%	20.0%	24.4%	27.3%	21.1%	24.4%	27.4%	21.5%	24.4%	26.8%	22.1%	24.4%	26.3%	22.7%	24.4%	26.0%
0.37	17.5%	24.2%	28.8%	19.9%	24.8%	28.5%	20.6%	25.1%	28.3%	21.9%	24.9%	27.2%	22.2%	25.1%	27.0%	23.0%	24.9%	26.8%
0.38	19.3%	25.1%	29.3%	20.4%	25.3%	29.1%	21.1%	25.4%	28.8%	22.3%	25.4%	27.9%	23.4%	25.5%	27.6%	23.6%	25.7%	27.5%
0.39	20.8%	25.9%	30.0%	20.8%	25.9%	29.1%	22.6%	26.3%	29.6%	23.5%	26.1%	28.5%	23.5%	26.2%	28.2%	24.4%	26.2%	28.3%
0.40	19.0%	26.4%	31.1%	19.4%	26.5%	30.2%	22.5%	26.5%	29.7%	23.3%	26.7%	29.8%	24.2%	26.8%	29.0%	24.6%	26.8%	28.5%
0.41	19.5%	27.4%	31.3%	23.1%	27.5%	31.4%	23.8%	27.4%	30.7%	23.8%	27.3%	30.0%	25.5%	27.6%	29.5%	25.3%	27.4%	29.2%
0.42	21.3%	27.6%	32.1%	22.3%	27.7%	31.6%	23.8%	28.0%	31.4%	25.0%	28.0%	31.1%	26.0%	28.1%	30.1%	25.9%	28.0%	29.9%
0.43	1.9%	11.2%	27.2%	22.2%	28.0%	32.6%	23.9%	28.3%	32.0%	25.3%	28.6%	31.5%	26.0%	28.4%	30.8%	26.5%	28.5%	30.5%
0.44	23.6%	29.1%	34.2%	24.4%	29.2%	33.3%	24.7%	28.9%	32.5%	25.6%	29.2%	32.3%	26.5%	29.1%	31.4%	26.9%	29.2%	31.3%
0.45	22.1%	29.0%	33.5%	24.8%	29.4%	34.2%	26.6%	30.1%	33.0%	26.3%	29.6%	32.6%	27.1%	29.7%	32.2%	27.1%	29.8%	31.9%
0.46	20.7%	29.7%	34.4%	23.9%	30.2%	34.4%	25.6%	29.9%	33.8%	27.2%	30.2%	33.4%	27.2%	30.3%	32.3%	27.8%	30.2%	32.2%
0.47	22.5%	30.5%	36.0%	24.2%	30.5%	35.2%	25.7%	30.7%	34.8%	27.4%	30.9%	34.3%	28.3%	30.9%	33.2%	28.7%	30.8%	33.0%
0.48	21.8%	31.0%	36.0%	25.4%	31.2%	36.2%	25.1%	31.2%	35.7%	27.9%	31.6%	34.6%	28.4%	31.3%	33.8%	29.5%	31.5%	33.7%
0.49	24.5%	31.7%	37.5%	27.2%	31.9%	36.1%	26.9%	31.6%	36.0%	28.9%	32.1%	34.8%	29.0%	32.0%	34.6%	29.9%	31.9%	33.6%
0.50	23.4%	31.9%	37.0%	25.7%	32.4%	36.8%	28.5%	32.4%	36.6%	28.3%	32.0%	35.0%	29.4%	32.6%	35.3%	30.4%	32.5%	34.7%
0.51	23.5%	32.4%	38.9%	26.8%	33.2%	37.5%	26.7%	32.5%	36.9%	29.6%	33.2%	35.8%	30.5%	33.4%	36.2%	30.8%	33.3%	35.4%
0.52	22.1%	33.0%	39.9%	25.7%	33.2%	38.0%	28.1%	33.3%	38.0%	30.6%	33.8%	36.9%	31.3%	33.9%	36.4%	31.3%	33.8%	35.6%
0.53	22.3%	33.4%	39.4%	28.6%	34.1%	39.5%	28.0%	34.3%	39.1%	29.7%	33.8%	37.1%	30.8%	34.2%	37.0%	31.5%	34.2%	36.4%
0.54	25.8%	33.5%	40.2%	29.1%	34.3%	39.7%	27.3%	34.3%	39.6%	29.8%	34.2%	37.8%	30.3%	34.9%	37.4%	32.3%	34.8%	37.1%
0.55	26.4%	35.5%	40.6%	31.0%	35.8%	40.7%	29.4%	35.5%	40.4%	31.9%	35.3%	38.3%	32.0%	35.2%	37.7%	31.9%	35.4%	37.5%
0.56	26.4%	35.8%	41.8%	27.8%	36.2%	41.4%	30.1%	35.6%	40.4%	31.5%	36.1%	40.0%	32.5%	35.9%	38.6%	33.6%	36.1%	37.8%
0.57	25.0%	35.7%	41.9%	28.6%	36.3%	41.2%	29.1%	35.9%	41.1%	32.3%	36.3%	40.7%	32.6%	36.4%	38.8%	34.1%	36.4%	38.5%
0.58	28.3%	37.0%	44.4%	27.8%	35.9%	41.6%	32.8%	36.9%	41.0%	32.8%	36.9%	40.1%	33.6%	36.8%	39.8%	34.5%	36.9%	39.1%
0.59	28.0%	36.9%	43.0%	29.0%	36.5%	43.2%	31.9%	36.8%	41.3%	32.1%	37.4%	40.8%	34.1%	37.3%	39.7%	33.8%	37.5%	39.9%
0.60	29.0%	38.1%	44.9%	31.2%	37.4%	41.7%	33.0%	38.1%	42.0%	35.0%	38.0%	40.7%	32.8%	38.2%	40.9%	35.0%	38.2%	40.7%

The forecasts for the next 15 days are then made using the seasonal and trend factors already identified by the team (Fig. 5.5). If resourcing is carried out using the forecast value, it can be seen that insufficient resources would be deployed on 7 days, failing to meet 5.5 % of the jobs arriving during the period.

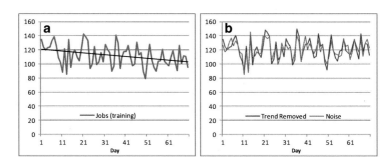

Fig. 5.4 Plot of tasks arriving at a work allocation team

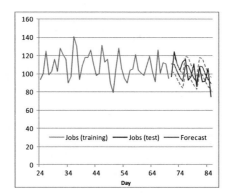

Fig. 5.5 Plot of tasks arriving at a work allocation team

By understanding that noise in the data will cause volumes of jobs to vary, and hence render the forecast inaccurate, resourcing can be carried out at a higher level (in this case, the forecast plus 8.9 %). In this scenario, the number of days where insufficient resources are deployed falls to 4 (out of 15) and just 2 % of the jobs remain unfinished.

5.6 Conclusion

It may appear unnecessary to know that there is a relationship between the amount of noise in a data set and the expected accuracy that is likely to be achieved from forecasting that data. For most purposes, the forecast accuracy ought to be sufficient to build in a degree of contingency into plans based on those forecasts. The relationship, however, does provide a number of benefits:

• To the forecaster—if a forecaster, whose task it is to identify the best forecast, is tasked with improving the forecasts, knowing (and being able to demonstrate) the limit of the data is invaluable.

- To the operational planner/manager—anyone who is responsible for turning the forecast into an operational plan will be able to understand the limitations of the forecast.
- To the performance manager—those responsible for setting targets (and those whose task is it to try and achieve them) will be able to establish sensible levels of performance objectives.
- To the process owner—knowing the level of the noise in the data enables decisions to be made on where to focus process improvement efforts, so that by reducing noise, improvements in forecasts may be achieved.
- To the scenario modeller—the modeller can make an assessment of the likely forecast accuracy based on a less noisy data set that may result from process improvements.

Ultimately, knowing the limits to forecast accuracy allows for a more conservative planning approach. There is little point in a planner waiting in expectation of forecasts of 95 % accuracy if the noise in the data is likely to limit that accuracy to between 85 % and 90 %.

This kind of result enables the planner to argue for a more flexible plan or for more resource or greater capacity.

References

Armstrong JS, Collopy F (1992) Error measures for generalizing about forecasting methods: empirical comparisons. Int J Forecast 8(1):69–80

Bunn DW, Taylor JW (2001) Setting accuracy targets for short-term judgemental sales forecasting. Int J Forecast 17(2):159–169

Decision 441 Forecasting (2005) Decision forecasting principles. http://www.duke.edu/~rnau/regnotes.htm

Hyndman RJ, Koehler AB (2006) Another look at measures of forecast accuracy. Int J Forecast 22 (4):679–688

KXEN (2012): Version 6.1.2 http://www.kxen.com/

Makridakis S, Wheelwright SC, Hyndman RJ (1998) Forecasting: methods and applications. Wiley, New York, NY

Malpass J, Shah M (2008) Forecasting and demand planning. In: Voudouris C, Owusu G, Dorne R, Lesaint D (eds) Service chain management. Springer, Heidelberg, pp 51–64

Minitab (2010) Version 16.1.0.0. http://www.minitab.com/en-GB/default.aspx

Oakland JS (1996) Statistical process control. Butterworth-Heinemann, Oxford

Signal to Noise Ratio (2008) In Wikipedia, The Free Encyclopedia. Retrieved February 26, 2013, from http://en.wikipedia.org/w/index.php?title=Signal_to_noise_ratio&oldid=204687150

SPSS (2010) Version 19.0.0. http://www-01.ibm.com/software/uk/analytics/spss/

Chapter 6
Modern Analytics in Field and Service Operations

Martin Spott, Detlef Nauck, and Paul Taylor

Abstract Businesses need to run their processes for field and service operations effectively and efficiently, providing good service at reasonable costs. Due to the changing nature of businesses including their environment and due to their intrinsic complexity, processes may require adaptation on a regular basis. Modern analytics can help improve processes and their execution by extracting the real process from workflow data (process mining), pointing to problems like bottlenecks and loops, by detecting emerging or changing patterns in demand and in the execution of processes (change pattern mining). We will present a variety of the tools and techniques we have designed covering the above and give examples for their successful application.

6.1 Introduction

With the industrial revolution, the introduction of mass production saw a drastic shift from artisanal crafting of products, usually performed by one person, to the large-scale production of goods, where each person was responsible for a single step in a process typically orchestrated by someone that was not involved in the execution. One of the first persons to define this series of steps as a formal process was Smith (1904), where he describes the series of steps required for the production of a pin.

With increasing demand for goods, the increase in competition and the need for a reduction in costs required the reengineering of the existing processes. In most major companies, this was already being considered even before it was formally defined during the 1990s by Hammer (1990) and Davenport and Short (1990). They introduced a procedural approach for improvement of business processes to ensure

M. Spott (✉) • D. Nauck • P. Taylor
Research and Innovation, BT Technology, Services and Operations, Martlesham Heath, UK

G. Owusu et al. (eds.), *Transforming Field and Service Operations*,
DOI 10.1007/978-3-642-44970-3_6, © Springer-Verlag Berlin Heidelberg 2013

correctness and to improve effectiveness, efficiency and compliance with statutes and protocols.

In the present day, business processes are well understood and formally defined with standardised representations and associated reengineering procedures. Business analysts exploit these formal representations by defining performance measures and quality constraints to analyse and improve specific aspects of the enterprise. Such business improvement activity is based on the process captured and defined using a formal language. However with the increase of the complexity of the formal languages used to model processes, we witness an increase in the distance between the formal process model and the process that is actually being executed (Browning 2009; Cardoso et al. 2009).

Large enterprises are required to actively respond to market demands and market evolution; therefore it is necessary to ensure control of crucial activities and to facilitate the reengineering of the processes running across the enterprise. Detecting evolution and change, be it external or internal, is however a non-trivial task. In many cases, change is only discovered when something has gone wrong which is particularly true for unexpected change. The question is if a machine can recognise trends automatically with only little guidance from a person.

This chapter presents a variety of the analytical techniques and tools we have designed covering various aspects of the above. In particular, Sect. 6.2 deals with business process mining, i.e. the extraction of business processes from data and techniques to find problems and automatic change detection. The sections include examples of successful applications and challenges we have encountered.

6.2 Business Process Mining

One way to exploit enterprise information is to reconstruct process execution in the organisation. This allows the behaviour of those various processes and the actors involved to be monitored in order to identify reasons for bottlenecks, incorrect executions, rewinds, loops and other issues preventing the process from matching the desired strategic requirements. However the process of extracting measurable evidence from the enterprise knowledge base is a non-trivial activity, which requires understanding and analysis of the activities and their interactions to transform them into measurable models.

6.2.1 The Aperture Process Mining Tool

A major issue in process analysis is to capture and interpret data correctly. There are languages such as BPMN (OMG 2011) and BPEL (OASIS 2007) that are used to explicitly define a process and capture execution information for fully automated systems (e.g. web service orchestration). However in most cases, integration of

BPEL engines in existing and ongoing processes is a difficult undertaking that enterprises prefer to avoid unless a minimum return of investment is guaranteed. This effort might also be undesirable where the process is not formally captured and it exists only in an idealised sense in the mind of the people involved. Therefore there are many situations where a process model is of very limited use or is simply not available; in which case, the process model needs to be inferred from the information created during process execution.

To respond to this specific requirement, we are using an internally developed process mining tool known as Aperture. It has been designed to provide an analysis that is focused not upon process mining itself but upon process analysis and business improvement. Aperture is able to reconstruct the process model from data stored across heterogeneous sources of information, providing the users with a unified framework to analyse processes and tasks that are executed across the enterprise, that otherwise would be extremely difficult to monitor and improve.

One of the design goals of Aperture is that it should be usable by those without experience or expertise in process mining—typically domain experts who understand the context of the processes. Traceability, that is the ability to trace the components of the generated model back to the source data, is therefore an important aspect which may help win over sceptics whose initial reaction is often to reject the mined model as being unrealistic or just plain wrong. Moreover the ability to trace each part back to the source builds confidence and leads to increased buy-in from the users.

Aperture's back-end data model consists of two main elements: the *process instance* and the *task instance*. Each *process instance* describes the execution of one job like the provision of an ordered service or the repair of a fault; each *process instance* consists of a number of *task instances* or *activities* linked together by a temporal relation (activity A is executed before activity B).

The minimum information required to create a process model to be used in Aperture is a process instance with a minimum of one task instance. The minimum amount of information to create a task instance is the start time and end time of each task instance. Process instance start time and end time can be derived from the starting time of the first task and the ending time of the last task if they are not explicitly provided.

The algorithm used by Aperture to connect tasks into a process instance is based upon minimum-spanning-tree algorithms (Eisner 1997). To ensure that the models are plausible, domain knowledge can be incorporated, thus placing appropriate constraints on the mining algorithm. An example of where this ability has been used is in a system handling complex orders with more than one suborder which are executed independently in parallel—this can sometimes be determined from the data but not in this case. Consequently, the process execution would not have been uniquely defined and default constraints in Aperture may have enforced process instances which do not reflect the real execution.

Our approach taken with Aperture differs significantly from that followed by other process mining tools such as ProM (van der Aalst et al. 2004) or DPM one from Pallas Athena. Their aim is to find the main process instances as a simple description of the real process to gain understanding at a higher level. To achieve this, they assume that only few process instances exist and interpret slight deviations from those as noise.

Aperture on the other hand is targeted at business process improvement which involves looking at potentially every process instance that may cause problems. Nevertheless, Aperture constructs the overall process as a compound model of the individual process instances. The following section describes various techniques in Aperture to analyse processes that can be used to detect points for improvement.

6.2.2 Process Analysis

The first and most intuitive of the analysis options offered by Aperture is the creation of the compound process model, as in the example in Fig. 6.1. This will combine the mined models for each process instance into a single graph showing the overall shape of the process. Weighting of the arcs of the graph in Fig. 6.1 is used to indicate frequency, i.e. how often the particular task transition has been executed over all process instances. In addition, the tool allows users to visually distinguish in the graph between the first time a task is executed in a process and where it is repeated; the first occurrence is connected using a blue (or solid) arc and the second using a red (or dashed) arc. In a compound diagram, this shows clearly where most of the repetition (or *rework*) is occurring and therefore where efficiency savings could potentially be made.

A set of more complex analysis tools can be used starting from the process model graph: drill-down facilities are provided on top of the model which allow the user to see, for example, if a particular problem transition is only associated with a particular group of process instances.

The second of the analysis options offered by Aperture is the extraction of the *common paths* through the process. A path is a distinct concrete ordering of tasks (either sequential or parallel) as executed in the source system(s); in Aperture the process instances that have the same concrete sequence are therefore considered to follow a common path. These can then be examined for frequency—how often they occur in relation to other paths—and for the commonality of their attributes, allowing correlations between attributes of the process and the execution path to be discovered.

Furthermore the tool provides ad hoc querying for Key Performance Indicators (KPIs) using either a custom workflow-oriented language called workflow execution language (WEL) (Majeed 2011) or more complex measures defined through the Groovy scripting language, interacting directly with the Aperture data model. In conjunction with the ability of the tool to extract subsets of data, this allows the user to quickly get an insight into the relative performance of those subsets in terms that are familiar from strategic reporting (once the formula for the KPIs has been entered into the tool).

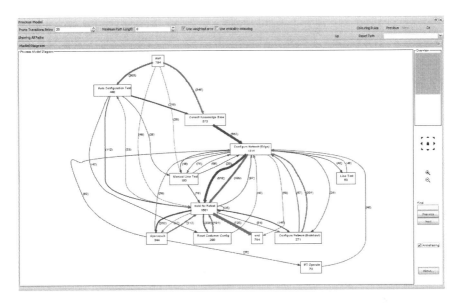

Fig. 6.1 A screenshot of aperture's main screen

6.2.3   Common Issues in Business Processes

We have applied Aperture to a variety of processes in a number of diverse multinational organisations, where we encountered the following generic issues and common problem patterns in the design of business processes, their implementation and execution:

- *Inadequate data model*
  Workflow data must be recorded in a way that represents the process sensibly for analysis, performance evaluation and improvement. Tasks need to be defined at a reasonable level of detail, time stamps should be taken for start time and end time of the actual execution such that delays between tasks can be detected. The combination of the above is also required to identify parallel tasks. We have come across task grouping in data that was masking the original workflow and therefore hindered any useful process analysis.
- *Inadequate data quality*
  As in all data mining tasks, adequate data quality is crucial for a successful analysis. We have encountered all common problems from missing to wrong values.
- *Variation in standardised processes*
  Processes are often standardised across different parts of a business; however different business units may run them differently. In one particular example, we found that the supposedly same process had been executed considerably less successfully in one country than in another one. One cause of problems was that

the assumptions for the standard process were met in one country, but not the other one. Aperture highlighted the problem and helped them improve the execution.

- *Design and implementation of processes not synchronised*
Processes may change during their lifetime and sometimes only the original version has been formally designed. In this case, any process improvement work needs to be based on the real process rather than the designed one which is more difficult because the documentation does not describe the real process sufficiently. We also found that some processes had been designed at a more abstract level than the implementation causing similar problems.

Aperture helps analysts identify all of the above issues; however, where variations in a process standard, for instance, can usually be resolved by adapting the running process model, problems like an inadequate data model are much deeper rooted in internal systems and require considerable effort for improvement.

6.3   Detecting Change Over Time

The design of business processes makes assumptions about the business itself, its infrastructure, the services it provides to customers and customer demand to name just a few. Modern approaches to process design and execution can adapt to change to a certain extent, but cannot deal with unexpected change. To consider all possible variations of the assumptions is obviously not feasible, but even to detect relevant change of assumptions in live processes is difficult. This goes back to an intrinsic problem of analytics: *we only find what we are looking for*. If we do not expect a particular change, we will not look for it and therefore either miss it or only detect it after it has caused problems. The following sections tackle this problem by introducing techniques that scan all available data and information for changes and present the most interesting findings to analysts. Such information can in turn be used to challenge the assumptions of running processes and it can be linked to performance measures of process, for instance, in order to explain performance issues.

6.3.1   From Hypothesis Testing to Generating Hypotheses

Traditionally, analytics means testing hypotheses. An analysis process consists of formulating a hypothesis based on domain knowledge and testing its validity. For instance, if the number of incoming jobs of a service provider is rising over time, then an analyst may test whether this was driven by an increasing demand for certain products, in certain locations, by specific types of job, etc. The number of hypotheses is however limited by the analysts' time and imagination. To make things worse, some analysts argue that they do not even know what they are looking for, that they will only know when they see it. This suggests that we must look for a clever way to find and present potentially interesting patterns to users rather than expect them to come up with hypotheses.

Unfortunately, machines still struggle to evaluate the interestingness of found patterns, to automatically make decisions and trigger actions based on them, mainly due to the lack of domain knowledge. For that reason, we are interested in extending exploratory data analysis in such a way that machines focus on the mechanical part of analysing large amounts of data and only support the analysts' exploration of the results through interactive visualisations. The analysts can then include domain knowledge, trigger further analysis by the machine and make decisions based on the results.

One core question is how to point an expert to the most relevant patterns. While it is very challenging to design an algorithmic method to assess the interestingness of a pattern, it is astonishingly simple for us humans to decide what is relevant to us and what is not. Though human decision-making mechanisms are not well understood, experience and domain knowledge of analysts that may not be available to a machine clearly contribute to human's advantage. Furthermore, based on our experience with analysts, trends and change of patterns over time are usually viewed as interesting.

In relation to this observation, there has been an increasing research interest over the last few years in methods which aim at analysing the changes within a domain by describing and modelling how the results of data mining—models and patterns—evolve over time. The term change mining has been coined as an umbrella term for such methods (Böttcher et al. 2008). Change mining approaches have been proposed for a variety of patterns and models, but many studies focus on analysing change in the context of item sets (market baskets or sets of associated attribute values), not only because item sets are rather comprehensible but also because their evolution can be represented in a convenient and interpretable way. We will therefore focus on such techniques in this chapter.

6.3.2 Application Example

For illustration, we will use the following exemplary problem throughout the following sections, which is very common across different industry sectors. Assume a service provider receives jobs such as orders or fault reports from customers. Every job has attributes such as time stamp (when the job came in or when it has been completed), type of job, location, type of customer, product and service level. Such attributes may differ slightly between industries, but at their core, they can be assumed to be typical.

The organisation needs to understand how the number of jobs develops over time for different attribute-value combinations such that new trends in particular segments can be spotted early. In this way, problems can be anticipated and resolved before they escalate and business processes may have to be adapted to such change. Furthermore, such trends can be used for forecasting and planning.

Most organisations monitor trends only at a global level. For instance, they would only look at the development of the overall number of jobs rather than all the different types of jobs in all different locations for all customer groups. Such an approach is reasonable, since the number of different attribute-value combinations grows exponentially with the number of attributes and can be very high—in some real data sets tens of thousands different combinations. If a high-level trend has been detected, the root cause needs to be found. Analysts will try to figure out if the trend is local, i.e. only for a certain type of job, product, location, etc. Since the number of different combinations is usually too high to test all of them manually, we are looking to automate the procedure and visualise the results for a guided exploration.

We assume that some attributes are hierarchical, i.e. they can be further broken down into attributes with a higher level of detail. Figure 6.2 shows—in partial—a hierarchy for the attribute *location*. Starting at UK level, each level below introduces an attribute which divides the country up into disjoint, more fine-grained parts. For *location* these attributes may be *region*, *county*, *borough* and so on, until *post code*. For now, we stipulate that every element in the hierarchy has exactly one parent element, i.e. hierarchies can be represented by trees. Additionally we require that the hierarchy is complete in the sense that it contains all children of a node. For instance, for *region* = *South*, we require all counties in the South of the UK as children. If a hierarchy is not complete, we can simply add a virtual value *other* covering for the missing values. Neither of the two assumptions is severe, since almost all hierarchies in our real-world data sets follow this schema.

Attributes with no hierarchy, i.e. just a flat set of values, can be given a value *all* as a virtual root node. *all* is equivalent to the set of all values, i.e. the attribute domain. In that way, every attribute can be made hierarchical for consistency.

As all attributes in data, such a hierarchy constrains the pattern detection process in so far as it defines the granularity of the data and the patterns. For instance, if there is an interesting development in job volumes in a particular geographical location that cannot be adequately described with the hierarchy above and if there is no finer-grained location information in the data, a machine will not be able to detect it. More broadly, data always comes with a point of view. This is one of the reasons why we advocate mixed-initiative approaches, where analysts contribute with their domain expertise and make sure that data suitably represents information.

In the following sections, two different approaches for detecting change patterns will be discussed. Both aim at supporting analysts to find the most interesting change patterns: one observes change at the highest level (the whole data set) and then guides an analyst towards contributing or deviating changes at lower levels, and one looks for interesting changes at low levels (large number of small subsets of the data) and aggregates them to a manageable number of changes for further analysis.

Fig. 6.2 Partial hierarchy of the attribute *location* as defined by the business

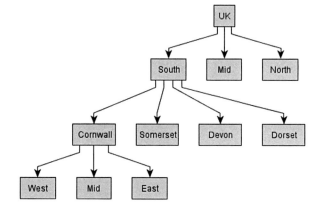

6.3.3   Drill Down from High-Level Trends: Top-Down Analysis

The idea of this exploration approach is to start with a high-level view of trends in the data, such as the overall trend of job volumes for the UK, and then guide the user in drilling down into attributes. For instance, if a trend is observed for the South of the UK and the same trend occurs in all its counties, a drill down into counties will not reveal more information. The other way around, if the trend prevalently occurs for a certain service, then the analyst should be made aware, eventually suggesting they drill down in this direction.

In order to decide whether drilling down into an attribute will reveal interesting patterns, we introduce the concept of *temporal homogeneity*. In general terms we consider a parent–child relationship in an attribute hierarchy temporally homogeneous, if the time series associated with each show the same trend over time. Derived from probability theory, *same trend* can be defined as a linear relationship between trends (Böttcher and Spott 2012). If a parent is temporally homogeneous with all its children there is no need to look at the trends of the children, we do not have to drill down.

In practice, testing for temporal homogeneity in terms of a perfect linear relationship between time series is too strict if the data is noisy. We suggest the use of measures to quantify the level of homogeneity or run statistical tests. Table 6.1 shows the Pearson correlation between the time series of a parent region (total) with the ones of the child regions. The time series describe the volume of jobs (weekly aggregated) over a period of 43 weeks in different regions (*location*).[1] The Pearson correlation is a measure of linear dependence, ± 1 meaning perfect linear dependence which is equivalent with temporal homogeneity according to our definition above. Region 6 shows the lowest level of correlation of 0.46 with the parent region (low homogeneity) and regions 4, 8 and 9 the highest value at 0.95.

---

[1] Real data from a telecommunications company.

Table 6.1 Normalised time series of job volumes in nine regions compared with the parent region (total)

Region	Time Series	Correlation
1		0.75
2		0.79
3		0.89
4		0.95
5		0.83
6		0.46
7		0.87
8		0.95
9		0.95
Total		1.00

Pearson correlation between parent and child region reflects the homogeneity of the time series

The normalised time series in Table 6.1 illustrate the difference in homogeneity between regions. The time series of regions with high correlation like region 9 match the one of the parent region (total) much better than the poorly correlated region 6.

The flaws of using the Pearson correlation as a measure for linear dependence are well known (e.g. its sensitivity to outliers), but it nevertheless is a useful measure in practice. Alternatively, statistical tests are described in Böttcher et al. (2009).

Rather than looking for temporal homogeneity as a measure of similarity, one may be interested in measures that help analysts identify the reason for upward or downward trends. For instance, if a strong upward trend in job volumes is detected at the top level, the business is interested in finding those subsets that contribute most to this overall upward trend. The attributes describing such subsets can be used by domain experts to constrain their search for the root cause of the overall upward trend.

Since high-level trends are usually monitored in the business (business process monitoring), developments of concern would be detected by operators and lead to further drill-down analyses. For this purpose, we have developed interactive visualisation techniques that make use of hierarchies and temporal homogeneity (Schmidt and Spott 2012). The idea is to visualise attributes with their hierarchies in a radial tree layout as shown in Fig. 6.3. Every attribute—in this case *location*, *product* and *service level*—is shown as a grey dot or triangle on the inner circle representing the root node of the attribute. Underlying hierarchies can be unfolded towards the outside of the radial tree as partially shown for *location* in Fig. 6.3.

Fig. 6.3 Radial tree layout
of attribute values, attribute
location partially unfolded

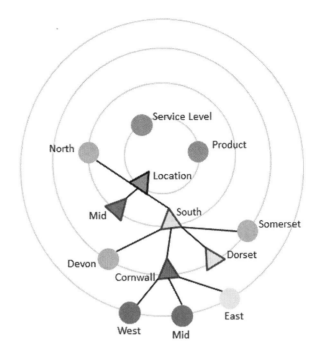

The shape of a node indicates if the underlying children are homogeneous (circle) or not (triangle). If not, the node can be unfolded by one level and the child nodes be revealed. Every child node has a colour and a shape: the colour to indicate its homogeneity with the parent (red, amber and green in Fig. 6.3) and the shape to indicate if the underlying children are homogeneous or not as before.

The hierarchies of all attributes can be opened up at the same time; however it must be noted that the homogeneity indicators in a hierarchy are always based on the assumption that the other attributes have the root value. In other words, we can only drill down into one attribute at a time. We can however select values lower down in all but the hierarchy for the drill down—which is equivalent to selecting a subset of the data as the new base line for homogeneity tests. The homogeneity indicators will then be recalculated and we can again drill down into the chosen attribute.

In a different analysis mode, homogeneity can be measured between a node and its root node rather than the immediate parent. Figure 6.4 shows a visualisation of this mode based on Mike Bostock's Sunburst diagram (Bostock 2012), implemented in d3. This example is essentially a visualisation of Table 6.1, but extended to the volume of jobs of nine regions over 32 weeks with a number of subregions each. The nine segments of the inner circle represent the regions, to which the segments of the associated subregions are attached in the outer circle. The grey levels indicate the level of homogeneity of a segment with the root, i.e. the homogeneity of the development of job volumes in the chosen segment (region or subregion) and the entire UK. Light grey means a high level of homogeneity,

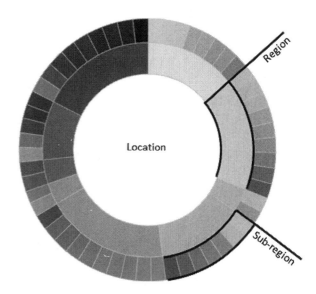

Fig. 6.4 Circular representation of homogeneity of nine regions (*inner circle*) and 54 subregions (*outer circle*) with the root. *Light grey* indicates a high level of homogeneity, *medium grey* a medium level and *black* a low level

medium grey a medium level and black a low level. An analyst will typically focus on the dark segments to look for unexpected developments over time. The sizes of the segments correspond to the volume of jobs in the different regions and help evaluate the gravity of differences. Labels can be added to the segments and sub-trees folded and unfolded. Furthermore, other attributes can be included in the circular representation.

The associated time series can be added to both the visualisation in Figs. 6.3 and 6.4 to compare the development over time in different subsets of the data.

6.3.4 Detect and Aggregate Low-Level Trends: Bottom-Up Analysis

The top-down technique described in Sect. 6.3.3 is useful for applications where analysts monitor how values develop over time at a high level and try to narrow down trends found at a global level to local ones. In general, interactive data exploration by drill down is typical for OLAP systems as part of corporate data warehouses, and the proposed approach augments such systems in that it offers a guided and more focused analysis.

While such a drill down is a useful tool for strategic and decision-making control on an upper management level, it has shortcomings at the operational level: first, a large number of attributes lead to an explosion in the number of paths an analyst may need to drill down into, which is not feasible. The discovered changes are thus still biased towards an analyst's preferences, and therefore changes may not be discovered at all. Second, the aforementioned hierarchies do not model strong dependencies between attributes values, as they occur when a certain service is only offered in a particular region but nowhere else. Since analysts have to drill down one attribute at a time, such dependencies are difficult to discover.

Both issues can be solved by analysing the time series and trends of every possible attribute-value combination. To deal with the combinatorial explosion, Böttcher et al. (2009) introduced a solution that tackles two main problems: first, extract, analyse and compare a large number of time series in reasonable time scales and, second, reduce the set of potentially interesting patterns to a size that can be managed by the analysts.

The first problem is solved by adapting *frequent item set discovery* introduced by Agrawal et al. (1993). The algorithm focuses on those attribute-value combinations that are frequent enough to be interesting. If adding an attribute value reduces the frequency to a value lower than a threshold, the algorithm will stop adding more attribute values. For instance, if the number of jobs related to product P, customer type C and location L is lower than the threshold, then adding information about an additional attribute like the service level will further reduce the number of jobs and is therefore unnecessary. This technique reduces the search space for exploration and at the same time the number of patterns, i.e. the second main problem from above.

Böttcher et al. (2009) developed two more techniques to reduce the number of patterns. The first removes redundant patterns based on temporal homogeneity as defined above. If removing an attribute from a pattern does not change the trend of its time series in terms of being temporally homogeneous, then the removed attribute does not contribute to the trend and can be left out. In this way, attribute-value combinations can be simplified as much as possible without losing information. Böttcher et al. (2009) describe three different criteria for temporally homogeneity, all based on probability theory. Furthermore, the authors measure the level of interestingness of patterns using time series analysis. The fact that a time series shows a trend over time like upward/downward trend, sudden change, spikes, etc. makes it more interesting than other ones. The strength of such trends can be measured and be used to rank the patterns such that analysts can filter for certain kinds of trends and then pick those patterns with the strongest examples.

Figure 6.5 shows a screenshot of our tool IDEAL that implements the techniques described above. The top panel lists the patterns found by IDEAL, represented as *association rules* or *item sets* (Agrawal et al. 1993), together with a few columns for measures of trend strength. The time series of the highlighted pattern are displayed in the bottom panel, i.e. the *support* (relative frequency of occurrence) and the *confidence* (relative frequency of the rule consequence given the antecedent) values over time.

Users find patterns with highly significant trends at the top of the sorted list and can then judge, if the found patterns are of interest. If that is the case, the trend of related patterns—patterns that share attribute values—can be compared with the first one in order to gain additional information. Furthermore, potentially unrelated patterns that show a related trend over time can be retrieved for the same reason.

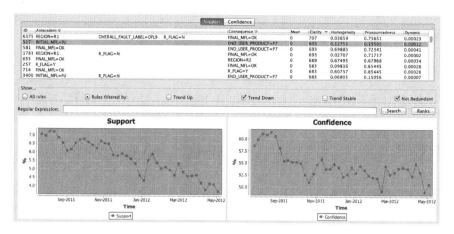

Fig. 6.5 *Screenshot* of our tool IDEAL that discovers temporal patterns from data and analyses and evaluates trends of the associated time series

6.4 Conclusions

Many problems in business processes can be attributed to changes within an organisation or to external customer behaviour and market evolution. Process designs are simply going out of sync with reality. This is only one of several reasons why many business processes are not executed as they have been designed. To manage this problem, we have developed the process mining and analysis tool Aperture that takes workflow data from operational systems and reconstructs and visualises processes as they are executed in operations. The result can be analysed in different ways, by looking at features like the variety of process executions, bottlenecks and loops in dependence of attributes like product, department and customer. Furthermore, performance metrics like duration and failure rate can be defined on process executions and can be used to find problematic execution paths.

Deterioration of business process performance is only a symptom of internal, organisational change or external developments. It is therefore important to detect such change directly and as early as possible rather than wait for businesses to perform poorly. Where high-level trends are usually tracked well, particularly gradual, granular changes are often missed, because nobody is looking for them and they are too subtle to be picked up at a higher level view.

IDEAL has been developed to pick up such changes and trends automatically. The two variants described in this chapter tackle the problem of spotting patterns from two angles, the first guiding users in a top-down approach from a high-level view towards lower level trends, and the second finds patterns at the lowest level, aggregates them for simplification if possible and ranks them according to their strength of trend.

Both Aperture and IDEAL have been successfully applied to reveal patterns in the organisation analysts were not aware of and they were consequently be able to improve related business processes.

Current and future work extends the presented projects in so far as we add predictive capabilities. For instance, we run experiments to see if the outcome of a process instance can be predicted well enough during its execution such that interventions can be triggered early to avoid problems. In IDEAL, detected trends in time series can be used to forecast future developments. As above, this allows analysts to detect problems early on before they escalate.

References

Agrawal R, Imielinski T, Swami A (1993) Mining association rules between sets of items in large databases. In: Proceedings of the ACM SIGMOD international conference on management of data. ACM, Washington, DC, pp 207–216

Bostock M (2012) Sunburst, visualisation example for d3. http://mbostock.github.com/d3/ex/sunburst.html

Böttcher M, Spott M (2012) Exploring time series of patterns: guided drill-down in hierarchies using change mining on frequent item sets. In: Moewes C et al (eds) Computational intelligence in intelligent data analysis studies in computational intelligence, vol 445. Springer, Heidelberg, pp 181–194

Böttcher M, Spiliopoulou M, Höppner F (2008) On exploiting the power of time in data mining. SIGKDD Explor 10(2):3–11

Böttcher M, Spott M, Kruse R (2009) A condensed representation of item sets for analysing their evolution over time. In: ECML PKDD 2009: Proceedings of the European conference on machine learning and knowledge discovery in databases. Lecture notes in artificial intelligence, vol 5781. Springer, Berlin, Heidelberg, pp 163–178

Browning TR (2009) On the alignment of the purposes and views of process models in project management. J Oper Manag 28(4):316–332

Cardoso J, Aalst W, Bussler C, Sheth A, Sandkuhl K (2009) Inter-enterprise system and application integration: a reality check. In: Filipe J, Cordeiro J, Cardoso J, Aalst W, Mylopoulos J, Rosemann M, Shaw MJ, Szyperski C (eds) Enterprise information systems, vol 12, Lecture notes in business information processing. Springer, Berlin, Heidelberg, pp 3–15

Davenport T, Short J (1990) The new industrial engineering: information technology and business process redesign. Sloan Manage Rev 31:11–27

Eisner J (1997) State-of-the-art algorithms for minimum spanning trees—a tutorial discussion. Master's thesis, University of Pennsylvania

Hammer M (1990) Reengineering work: don't automate, obliterate. Har Bus Rev 68(4):104–112

Majeed B (2011) Us patent number 2011/0093308 a1. Process monitoring system

OASIS (2007) Web services business process execution language version 2.0 (April)

OMG (2011) Business process model and notation (bpmn) version 2.0 (January)

Schmidt F, Spott M (2012) Visualising temporal item sets—guided drill-down with hierarchical attributes. In: Proceedings of soft methods in probability and statistics (SMPS 2012)

Smith A (1904) An inquiry into the nature and causes of the wealth of nations, 5th edn. Methuen & Co Ltd, London, Republished from: Edwin cannan's annotated edition, 1904 edn

van der Aalst WMP, Weijters T, Maruster L (2004) Workflow mining: discovering process models from event logs. IEEE Trans Knowl Data Eng 16(9):1128–1142

Chapter 7
Enhancing Field Service Operations via Fuzzy Automation of Tactical Supply Plan

Sid Shakya, Summer Kassem, Ahmed Mohamed, Hani Hagras, and Gilbert Owusu

Abstract Tactical supply planning (TSP) is an integral part of the end-to-end field resource planning process. It takes as input, constrained demand from the strategic plan at monthly (or quarterly) level, decomposes it to daily or weekly level and plans the capacity accordingly to meet the expected demand. The plan is then executed and sent to a work allocation system for on-the-day scheduling of individuals tasks to resources. A tactical supply plan ensures that there are enough resources available in the field on any given day. It highlights underutilised resources and offers recommendations on how best to deploy surplus resources. As such, TSP focuses on improving customer satisfaction by minimising operational cost and maximising right-first-time (RFT) objectives.

In this chapter, we describe opportunities and challenges in automating tactical supply planning and present a fuzzy approach to address the challenges. The motivation is to minimise the effort required for producing a resource plan. More importantly, our objective is to leverage computation intelligence to produce optimised supply plan in order to increase RFT and the customer satisfaction.

7.1 Introduction

It is well recognised that one of the key contributors to the success of any firm is the proactive planning of its resources to meet the expected demand. This is crucial for service industries, especially for those with large and dynamic workforces. For them, advance planning of their resources largely determines the cost and quality of

S. Shakya (✉) • G. Owusu
Research and Innovation, BT Technology, Services and Operations, Martlesham Heath, UK
e-mail: sid.shakya@bt.com

S. Kassem • A. Mohamed • H. Hagras
School of Computer Science and Electronic Engineering, University of Essex, Essex, UK

G. Owusu et al. (eds.), *Transforming Field and Service Operations*,
DOI 10.1007/978-3-642-44970-3_7, © Springer-Verlag Berlin Heidelberg 2013

their services and also their customer satisfaction. These measures also dictate their revenue.

Tactical supply planning (TSP) is an important part of end-to-end resource planning process. It sits between long-term strategic supply planning and the short-term scheduling (Fig. 7.1). Strategic supply planning is typically done for the period of 12–18 months. It looks at expected unconstrained demand for its product and services, typically at monthly level, combines it with different business objectives and makes decisions on how much capacity is to be made available in order to best match the demand and at the same time adhere to various business and financial constraints. The output of this process is the constrained high-level demand for products and services that the business has committed to deliver based on constrained capacity. This is then fed to tactical supply planning which is typically preformed for 1–3 months. It takes constrained demand for products and services at monthly (or quarterly) level, applies various rules to decompose product into individual activities and produces daily or weekly demand at activity level. It also takes any actual demand that is already visible to the firm and combines it with the forecasted demand. These fine-grained demand profiles at activity level are then compared to the available supply and their skills, to make sure that there is enough supply to match the expected demand. At this stage, resource planners have to take into account any shortage (or surplus) in available capacity and optimise the plan so that, on the day, there is enough supply available to execute the tasks.

The output of tactical capacity planning goes to both the reservation system and the scheduling system. Reservation system uses it to make sure that booking for jobs are taken not exceeding the planned capacity. Similarly, the shorter term planning for 1–7 days is used by scheduler where individuals are scheduled according to their planned geography and skills, so that the utilisation of the individuals are maximised and the cost related to their idle time and travel are minimised.

It is important to note that, on one hand, tactical supply plan ensures that there are enough resources available in the field for scheduler to schedule work properly. On the other hand, it also makes sure that there are no resources left unused. As such, TSP contributes heavily to minimising operational cost and at the same time maximising RFT, leading to improved customer satisfaction.

In this chapter, we highlight some of the opportunities and challenges in TSP and present a fuzzy logic-based approach to solve the TSP problem. The motivation is to minimise the effort and the cost of producing plan. We demonstrate how we exploit computation intelligence technologies to produce optimised supply plan in order to increase RFT and customer satisfaction.

This chapter is divided into five sections. Section 7.2 describes the current manual TSP process. Section 7.3 highlights some of the key benefits of its automation. Section 7.4 describes some of the key trends in automating TSP process. Section 7.5 presents a fuzzy approach to TSP. Finally Sect. 7.6 concludes the chapter.

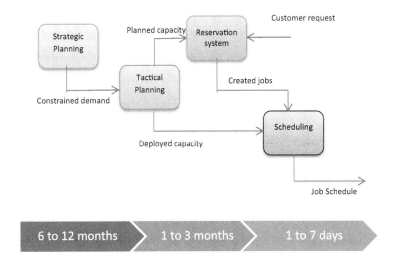

Fig. 7.1 Three stages of resource management

7.2 Current Approaches to TSP

It is well observed that the focus of the service industries mainly revolves around two aspects of supply chain management:

1. Strategic Planning—mainly due to the fact that it underpins finance and budgets which is crucial part of the business planning process
2. Scheduling and Execution—mainly to make sure that the tasks are executed and the customer's request is satisfied on the day

However, despite its importance in the success of end-to-end delivery of services, TSP is often overlooked and poorly managed. This is mainly due to the lack of expertise and the lack of tools and technologies for realising TSP. Current approach to TSP is mainly manual, executed by field resource planners, which involves making informed decisions on the movement of resources between different geographies and skills. This gets complicated with any increase in the number of resources and the types of products and services that a firm offers, consequently the increasing number of skills/areas and the number of plans that has to be balanced. Clearly, this is an optimisation problem, where the goal is to find the best moves that optimise the plan across all skills, geography and time dimension.

In many cases, companies tend to react to capacity imbalances, by bringing in people on overtime or bringing in contingency resources. This reactive approach can lead to inefficiencies and potential failure of customer commitments. Furthermore, even when capacity is managed proactively, it is mainly performed manually, maintained in spreadsheets and using the manual decisions made by the resource planners. For organisations with large number of multi-skilled resources and with many different products and services, such manual decisions are (a) costly and (b) unlikely to be optimal. This clearly highlights the need for an automated approach.

7.3 Benefits of TSP Automation

Below we highlight some of the key benefits of automating TSP process.

7.3.1 Optimal Planning

As mentioned earlier, tactical supply planning can be seen as an optimisation problem. Let's look at a typical scenario when there is a capacity shortage for a particular skill and planner has to plan for the shortage. It is obvious for him/her to look for surplus resources in other plans. Also, resources can only be moved if they are compatible, i.e. if they have secondary skills to do the work in shortage plan. This task of moving resource is trivial when the number of plans is small and resources have only one or two skills. However, when there are multiple plans and resources are highly multi-skilled, possible options for resource movement become large. Let us illustrate this with the figures below:

Here, skills in red are the skills with capacity shortage, skills in blue are with surplus, and skills in green have no surplus or shortages, i.e. they are balanced. Figure 7.2a shows that Skill S2 and Skill S3 have shortages. Skill S2 can take from both Skill S4 and Skill S1 but Skill S3 can only receive capacity from Skill S1. Thus, before making any movement decisions, all of the skills and their dependent skills have to be evaluated. This becomes very complex when many as skills as geographies are involved. Figure 7.2b shows another complication, where Skill S3 can only receive capacity from Skill S2 but Skill S2 does not have any surplus; however Skill S1 or Skill S4, both with Surplus, can give to Skill S2. Here the solution would be to move resource from Skill S2 to Skill S3 and then move resource from Skill S1 or Skill S4 to Skill S2. Obviously, as the number of skills increases, such relationship can become very complex. In such cases any manual movement is unlikely to be optimal. An automated process is therefore required which can intelligently allocate resources and best optimise the plan.

7.3.2 Rapid Scenario Modelling

In most businesses, there are always priorities associated with specific skills (or areas), i.e. there could be skills that can be more business critical than others. There could also be different cost associated with different skills. Similarly, product

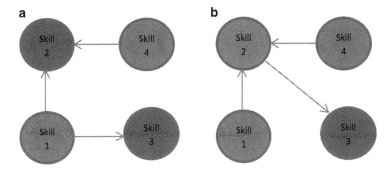

Fig. 7.2 Dependency in multi-skill planning. *Red* colours are shortage skills, *blue* are surplus skills, and *green* are balanced skills. The *arrow* signifies compatibility

offerings may also have different levels of SLAs. And in the situations where there are not enough resources to fulfil the demand for all skills or areas, operational managers will find the best option to utilise their limited supply. They may want to analyse changes in cost and SLA and their effect on resourcing. There could be other multiple 'what-if' scenarios that business may want to evaluate, before making any decisions on supply distribution. Manually setting up these scenarios and testing them could be very time consuming and likely to be suboptimal. Automation of TSP could be very useful for quick scenario modelling and evaluating different supply planning options.

7.3.3 Cost-Effective

As stated above, manual planning process can be time consuming and therefore can be very expensive. Automation can significantly reduce the cost of planning. Further more effective planning leads to effective execution of work and reduction in repeat visits to customers, which in turn delivers cost savings.

7.3.4 Human Error Elimination

Manual processes are prone to human error. Moving non-compatible resources can not only result in capacity waste but also increase the possibility of SLA failure. Clearly, automation using rules would eliminate such errors.

7.3.5   Improved Customer Satisfaction

Also, effective planning means effective and timely delivery of service, which increases customer satisfaction and implicitly helps generate revenue.

7.4   Current Automation: Trends

One of the simplest ways to automate TSP is by using a rule-based systems. They are extensively used in designing and manufacturing industries. They are simple to understand, are easier to build and easier to maintain and can be very effective if specified accurately. The idea is to imitate the human decision-making process by codifying and executing the rules they use in making decisions.

The key properties of any rule-based automation system for TSP are:

1. To allow the user to maintain multiple sets of rules for modelling different scenarios
2. To locate the shortages
3. To locate the surpluses
4. To balance the plan by executing the selected set of rules

Figure 7.3 shows an interface of a rule builder in FieldPlan (Kern et al. 2009; Lesaint et al. 1997, 2000), a resource planning component of Field Optimisation Suite (FOS).

The rules mainly specify the similarities between the skills (or areas) and specify how the system should behave when it encounters a shortage or a surplus, mainly by allowing series of moves to be performed to balance the plan. This approach works well when the relationships between skills (or areas) are simpler. It can also significantly reduce the time needed to perform the planning. However, it is likely that rules defined may not cover all possible shuffles that can arise within specific scenario and hence need for a system that tries to find alternatives rather than those specified in the rule set.

Some of the other sophisticated approaches to TSP include heuristic and meta-heuristic search methods including the Hill Climber algorithm (HCL), the Fast Local Search (FLS), Guided Local Search (GLS) algorithm, Simulated Annealing and Tabu search (Merz and Freisleben 1999; Bianchi et al. 2004; Dorne and Voudouris 2001; Wang et al. 2006). These methods model the TSP problem as an optimisation problem and assign a cost to each planning solution. The goal of these algorithms is to find the solution with the minimal cost. The simplest in this class of algorithms are Hill Climber algorithms. They are simple to implement and have easier workflows. One of the Hill Climber algorithms that was shown to give the best results employs the following steps (Dorne and Voudouris 2001; Shakya 2004):

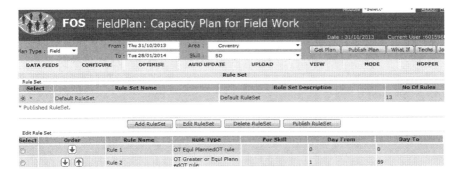

Fig. 7.3 A planning rule creation interface in FieldPlan system

1. Construct an initial solution which is called the baseline plan.
2. Gradually modify and improve the solution until a stopping condition (threshold of cost function value reached, running time expired) is satisfied by:

    (a) Generating the set of moves of resources across possible plans
    (b) Selecting the best move (the one giving the lowest value of cost function) and applying on the current solution

The efficiency of the meta-heuristic algorithm relies on the way the filtering of moves is guided across the search space, that is, the drivers that are used to filter the resource to move and to which plan to assign them (Dorne and Voudouris 2001). Similarly their effectiveness relies on the way the cost function is modelled. It is sometimes very difficult to correctly model the cost function where there are multiple objectives for optimising plan, e.g. minimise the travel of the resources, and at the same time covering the area and skills with higher priority. Using the crisp boundary in defining the cost function (e.g. for defining importance score for area and skill, and maximum travel limit) may not be effective.

In the next section, we describe a TSP planning approach to overcome some of the limitations of crisp algorithms as well as traditional rule-based systems. It is based on fuzzy logic. This approach allows flexibility in defining cost functions and allows greater flexibility in optimising plan.

7.5 Fuzzy Logic-Based TSP

Fuzzy logic is a well-known technique in AI that is use to model the situation where the decision-making logic is not crisp (Hagras 2004; Mendel 2001; Kassem et al. 2012). They are credited with providing transparent methodologies that can deal with the imprecision and uncertainties. For example, a crisp logic clearly distinguishes the difference between a tall and a short person by assuming a crisp point of differentiation, e.g. anyone above 160 cm of height is a tall person.

However fuzzy logic-based model does not have such cut-off point. Instead, it provides a confidence level that says if a person is 155 cm, he/she is short with a certain degree of confidence and tall with a certain degree. And therefore, a height of 155 cm can contribute to both tall and short values for the height variable. This can have different effect to a fuzzy cost function than in a crisp cost function.

7.5.1   Problem Formulation

The key idea for solving TSP is to model the planning process as a fuzzy decision-making process. For the purpose of this chapter, we simplify the problem to skill-based planning; however the method is easily generalised to include both skill- and area-based planning.

In the proposed method, we define two key parameters that formulate TSP problem:

- *Skill priorities.* For every skill a priority is assigned. This is to model the importance of each skill and is used to prioritise demand fulfilment. For example, priorities can be defined by one of Very Low (VL), Low (L), Medium (M), High (H) or Very High (VH). Skill priority is essential as in some cases there could be shortages in more than one skill, and the available surplus might not be enough, in which case, the higher priority skills should be covered first.
- *Skill compatibilities.* For every skill, a list of compatible skills is maintained. This models resources that have more than one skill and therefore can work in any of those skills as required. Also for every compatible skill, a percentage of compatibility is recorded. This percentage represents the percentage of resources that have the main skill but are still capable of covering the compatible skills. This is useful for the case where not all resources have skill to do other jobs.

The information on skill priorities and compatibilities can be represented using the graph shown in Fig. 7.4. This is an example of the relations between five skills. Each skill is represented by a square that contains the name of the skill (top of the square) and its priority (bottom of the square). An arrow from one skill to another means that these two skills are compatible. Here, the compatibility percentage is defined by a number on the arrow.

The above information imposes a certain order in performing the moves from one skill to another. For example:

1. It is quite intuitive that the highest priority skill should be addressed first. Then a surplus in any of the skills in the list of compatible skills should be searched for in order to balance the shortage.
2. It is also quite intuitive that the list of compatible skills will be searched in descending order according to the compatibility percentage.

However, this imposed order is not necessarily the order that will balance out the plan, as a skill could be compatible with multiple other skills. Being highly

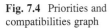
Fig. 7.4 Priorities and
compatibilities graph

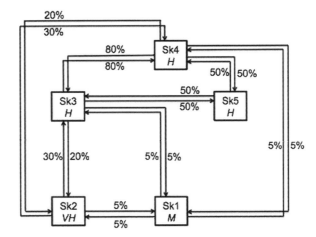

compatible may not mean the best one to take from, as other skills may have low
compatibility but could be scarce skill. Therefore, giving capacity to such scarce
skill first may better optimise the plan. This also means searching for all of the
linkages before making decision, which is computationally very expensive. This
highlights the need for a logic-based reasoning system that can handle the encountered uncertainties.

The aim of the fuzzy logic system (FLS)-based TSP is to find the best order to
perform the set of moves implied by priority and compatibility information.

7.5.2   Methodology

In this section, we will explain the core components of a FLS planning:

7.5.2.1   Parameter Calculation

There are three main parameters that the fuzzy system calculates. All of these
parameters are calculated for every possible move for each skill based on compatibility as follows

(a) Highest Surplus: The highest surplus is the highest man hours (MH) of
    resources that can be given from the source skill to the destination skill. This
    value is calculated by finding the surplus that the source skill would give if the
    order imposed by the compatibilities is followed.
(b) Lowest Surplus: The lowest surplus is the least value that is needed from the
    source skill. This value is calculated by neglecting this skill and finding how
    much of the shortage would be covered if all the other skills were exhausted. At
    the end of a processing cycle, if there is still a shortage, then this is the least

value that has to be given by the source skill. For example, if Skill S1 has the option to cover its shortage from skills S2, S3 and S4, where S1 has a shortage of 20 MHs, the surplus skills (S2, S3 and S4) have 10, 10 and 5 MHs, respectively. When we attempt to find the lowest value for S3, we first aim to cover the shortage using the other skills, which means that 15 MHs will be covered (from S2 and S4) and then the lowest value that S3 can give is 5 MHs.

(c) Preference: The preference is a new order that overrides the compatibility percentage, so rather than using this percentage, the compatible skills for one destination skill are ordered in ascending order based on which skill is least needed by other skills.

7.5.2.2 The Type-2 FLS Operation

The FLS processes one move at a time and determines the amount of MHs that will be moved.

(a) Inputs: For every move the FLS takes the highest surplus value, the lowest surplus value and the summation of all the lowest surpluses that are related to the source skill, as input. The third input represents how much the other skills need this surplus. This input is calculated by checking every move in the list of generated moves, if the source skill in that move is the same one as the one being currently processed, the lowest surplus value for that surplus is added to the summation and the final value of the summation is passed to the FLS. Every one of these three inputs is represented using a fuzzy variable that consists of five fuzzy sets: {None, Low, Medium, High, All}. An example of one of the inputs can be shown in Fig. 7.5. The thick lines represent the type-2 fuzzy sets (Hagras 2004) and the dotted lines s represent the type-1 fuzzy sets.

The fuzzy sets *None* and *All* are singletons, and the rest are type-2 fuzzy sets. Each of the type-2 fuzzy sets Low, Medium and High has a certain number of parameters, for example, the set Low has four parameters, the set Medium has five, and High has four. The universe of discourse is divided equally among the total number of parameters in order to determine the area covered by each set.

As shown in Fig. 7.5, for every fuzzy input there is a universe of discourse (the range of values on the x-axis). For the highest surplus value input, the universe of discourse starts from zero and is bounded by the shortage in the destination skill (MHs). The universe of discourse for the lowest surplus value is also bounded by the shortage in the destination skill. For the third input, 'the summation of lowest', the universe of discourse in this case is bounded by the surplus of the source skill.

(b) For every move processed by the FLS, there is a source skill and a destination skill. The destination skill is the one that has the shortage. The source skill is the one that should have the surplus that will cover the shortage. In some cases, the destination skill may have a shortage; however it may have more than one option in order to cover that shortage, i.e. more than one surplus skill. So the main aim of the inference engine is to determine whether this destination skill is

Fig. 7.5 Type-1 and type-2 fuzzy sets representing the highest surplus input

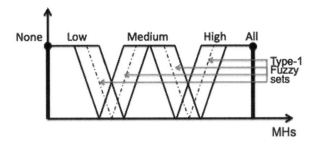

the skill that needs the surplus the most. In the case where the destination skill is not the skill that needs this surplus the most, then the inference determines how much of the surplus this skill takes in order to make sure that there is still surplus left for the other skills (i.e. that need the surplus more).

The first two inputs entered to the FLS (highest surplus and lowest surplus) are supposed to determine how much the destination skill needs the surplus in the source skill. For example, if the highest and the lowest values are quite close, this means that the value in the lowest surplus is absolutely indispensable. However as the value in the lowest surplus approaches zero and the difference between the lowest and the highest grows, this means that the surplus is not really essential to the destination skill—it probably has other options to cover its shortage. We will provide an example to illustrate the above scenario. If we have a destination skill S1 that needs 30 MHs, we have source skills S2, S3 and S4 where each has a surplus of 30 MHs. In this case the highest value for all three will be 30 MHs and the lowest value for all three will be 0 MHs. As the difference between the lowest and the highest is quite big, and the fact that the lowest is 0 MHs, we can get indication that the destination skill has more than one option.

The third input, which is the summation of lowest values, is supposed to represent all the lowest values related to the source skill. So these are the lowest values for all the other destination skills that use this same source skill. This represents how much this source skill is needed by other destination skills. This insight can be used to determine how much of the surplus of this skill can we take in this move, before affecting the other shortages.

After this step, the system determines how much the destination skills need the surplus and how much the other skills need this surplus. The system then uses its rule base to infer how much should be moved.

7.5.2.3   Rule Base

The rule base for the proposed type-2 FLS contains a number of rules. Here are some examples of rules and the underlying logic.

IF Highest Surplus is LOW AND Lowest surplus is NONE AND Lowest Summation
    is ALL
THEN move NONE

The logic underpinning this rule is—if the highest surplus value that the source skill can give is LOW, and the lowest surplus value that the source skill can give is NONE, the surplus provided by this skill is not only a small amount but it is also dispensable because the lowest value is NONE. This means that this shortage can be covered by other skills. The last input which is the summation of lowest is ALL means that there are a lot of other destination skills that need this surplus. Hence the value to be moved from this surplus is NONE, since this destination skill does not need this surplus and there are other skills that do need it.

Another rule example is as follows:

IF Highest Surplus is MEDIUM AND Lowest Surplus is LOW AND LowestSummation is LOW
THEN move MEDIUM

This rule means if the highest value that the source skill can give is MEDIUM and the lowest value that the source skill can give is LOW, then this destination skill can at least take a LOW percentage of the surplus and at most needs a MEDIUM percentage. The summation of lowest is LOW means that it is ok to take the highest surplus value as there are not many skills that need this surplus. So the value to be moved is MEDIUM.

Another rule example is as follows:

IF Highest is ALL AND Lowest is ALL AND LowestSummation is ALL
THEN move ALL

This rule means if the highest value that the source skill can give is ALL (which means its entire surplus) and the lowest value that the source skill can give is also ALL, then the entire surplus is absolutely necessary to this destination skill and it has no other option. In this case it does not really matter what the summation of lowest is because either way the entire surplus has to be taken by this destination skill.

7.5.2.4 Output

The output of the FLS is the amount to move from each skill to other compatible skills, which is also represented by a type-2 fuzzy variable using five fuzzy sets {None, Low, Medium, High, All}. The universe of discourse is the surplus of the source skill. This suggested move aims to define the optimal movement in order to balance the plan and best utilise the available resources.

7.6 Conclusions

The approach was tested with a field force planning scenario within BT. BT has more than 23,000 field and desk-based technicians. First, these human resources have to be carefully managed and balanced, so that on any given day, there is no shortage or surplus of capacity. Second, they have to be deployed so as to maximise their productivity. BT uses FOS (Field Optimisation Suite) system for automated resource management. FOS consists of a suite of applications that provides end-to-end resource management capabilities, starting from demand forecasting, capacity planning and capacity reservation down to deployment planning and resource scheduling capabilities for the execution day.

A case study was conducted within one of the BT's lines of business. The service area was related to installation and maintenance of telecommunication services. The nature of the work meant they were mainly short duration tasks requiring less than 2 h to complete. The forecasting component of FOS was used to forecast demand volumes. A rule-base system was used to dictate the movement of capacity to match the forecasted demand. The move made by rule-based system was compared against the moves suggested by the fuzzy TSP system. The key measure for comparison was the total collective shortage and surplus MH of capacity against demand for all areas and skills. The closer this value was to 0 the better the performance was—it indicated that the automated planning was able to achieve a balanced plan (i.e. the number of shortage and surplus hours was minimised). The results showed that fuzzy planning approach was able to improve resource balancing in the capacity plan by 6 % in comparison to the rule-based TSP system (Kassem et al. 2012).

Increasingly, organisations are automating their resource management processes. Unfortunately, very little attention has been given to automating the tactical supply planning process (TSP). Automating the TSP process offers opportunities to realise cost minimisation and revenue generation. In this chapter, we have reviewed several different ways to automate TSP process and presented in detail a fuzzy logic-based approach for TSP. The key differentiator here is its ability to smooth the effect of a variable on a cost function and therefore on the overall quality of the plan. The choice of TSP techniques depends on the requirements and size of the organisation.

References

Bianchi L, Birattari M, Chiarandini M, Manfrin M, Mastrolilli M, Paquete L, Rossi-Doria O, Schiavinotto T (2004) Metaheuristics for the vehicle routing problem with stochastic demands. In: Proceedings of eighth international conference on parallel problem solving from nature. Lecture notes in computer science, vol 3242. Springer, Berlin, pp 450–460

Dorne R, Voudouris C (2001) HSF: a generic framework to easily design meta-heuristic methods. In: Proceedings of MIC 4th metaheuristics international conference, Porto, Portugal, 16–20 July. pp 423–427

Hagras H (2004) A hierarchical type-2 fuzzy logic control architecture for autonomous mobile robots. IEEE Trans Fuzzy Syst 12(4):524–539

Kassem S, Hagras H, Owusu G, Shakya S (2012) A type2 fuzzy logic system for workforce management in the telecommunications domain. In: 2012 I.E. international conference on fuzzy systems (FUZZ-IEEE), 10–15 June. pp 1, 8

Kern M, Shakya S, Owusu G (2009) Integrated resource planning for diverse workforces. Computers and industrial engineering. In: Proceedings of the IEEE international conference on computers and industrial engineering, Troyes. pp 1169–1173

Lesaint D, Voudouris C, Azarmi N, Laithwaite B (1997) Dynamic workforce management. In: Proceeding of the IEE colloquium on AI for network management systems, vol 1. London, UK, pp 1–5

Lesaint D, Voudouris C, Azarmi N (2000) Dynamic workforce scheduling for British telecommunications plc. Interfaces 30(1):45–56

Mendel J (2001) Uncertain rule-based fuzzy logic systems: introduction and new directions. Prentice-Hall, Upper Saddle River, NJ

Merz P, Freisleben B (1999) A comparison of memetic algorithms, tabu search and ant colonies for the quadratic assignment problem. In: Proceedings of the 1999 congress on evolutionary computation, Washington, DC

Shakya SK (2004) Markov random field modeling of genetic algorithms. In: Progress report submitted to The Robert Gordon University to make the case for transfer from MPhil to PhD, The Robert Gordon University

Wang D, Zeng X, Keane J (2006) A survey of hierarchical fuzzy systems. Int J Comput Cognit 4 (1):18–28

Part III
Case Studies

Chapter 8
The Role of Search for Field Force Knowledge Management

Dyaa Albakour, Géry Ducatel, and Udo Kruschwitz

Abstract Search has become a ubiquitous, everyday activity, but finding the right information at the right time in an electronic document collection can still be a very challenging process. Significant time is being spent on identifying suitable search terms, exploring matching documents, rephrasing the search request and assessing whether a document contains the information sought. Once another user is faced with a similar information need, the whole process starts again. There is significant potential in cutting down on this activity by taking a user straight to the required information. As well as delivering technical information and vital regulatory information, a knowledge management solution is concerned with capturing valuable insight and experience in order to share it amongst workers. A search engine has been developed and deployed to technical support staff and we were able to assess its impact on mobile workers. The architecture is based on open-source software to satisfy the basic search functionality, such as indexing, search result ranking, faceting and spell checking. The search engine indexes a number of knowledge repositories relevant to the field engineers. On top of that we have developed an *adaptive* query suggestion mechanism called Sunny Aberdeen. *Query suggestions* provide an interactive feature that can guide the user through the search process by providing alternative terminology or suggesting 'best matches'. In our search engine, the query suggestions are generated and adapted over time using state-of-the-art machine learning approaches, which exploit past user interactions with the search engine to derive query suggestions. Apart from continuously updating the suggestions, this framework is also capable of reflecting current search trends as well as forgetting relations that are no longer relevant. Query log analysis

D. Albakour (✉)
School of Computer Science, University of Glasgow, Glasgow, UK
e-mail: dyaa.albakour@glasgow.ac.uk

G. Ducatel
Research and Innovation, BT Technology, Services and Operations, Martlesham Heath, UK

U. Kruschwitz
School of Computer Science and Electronic Engineering, University of Essex, Essex, UK

G. Owusu et al. (eds.), *Transforming Field and Service Operations*,
DOI 10.1007/978-3-642-44970-3_8, © Springer-Verlag Berlin Heidelberg 2013

of the system running in a real-life context indicates that the system was able to cut down the number of repeat faults and speeds up the decision process for sending out staff to certain jobs.

8.1 Introduction

Mobile IT technology is changing access to information for field workers—high-quality client devices can be carried whilst working away from office buildings. This opens the door to innovative knowledge management applications for mobile workers. Integration of mobile devices with enterprise networks is becoming more effective, with software-only solutions available and increased security. However, strategic knowledge management solutions are very costly and therefore changes are likely to be prudent. Pan enterprise solutions which store, distribute or capture information from field workers are not yet available, and the same is true for cloud solutions.

Information on enterprise networks is often stored in lengthy documents, across different silos with complex access rights. Access to this information is designed for desktop workers who can use tools to view, browse and even modify information. With regard to search engines, users tend to show some tolerance with regard to inaccuracy, given that query refinement is available; more complex interfaces (including facets, word clouds, filtering, etc.) have become the norm, but again, this is not something that is well adapted to mobility.

This chapter looks at field workers in particular and presents a knowledge management solution for accessing information. This solution, as seen in Fig. 8.1, involves a support line where office workers are in charge of finding information for field workers and deciding on a line of action varying from verbal advice, electronic information sharing or even team support. Access to information is basically delegated to office workers and communicated back using whichever means is appropriate. This pilot scheme was also recording incidents to build a knowledge base, but this is outside the scope of the work presented in this chapter. The pilot scheme ran with an approximate ratio of support staff to field engineers of 1–80.

The support line has been provided with a search engine dedicated to all relevant technical repositories. The search engine solution technical features are described in the following sections. The solution is not only relevant from a technical point of view; the software design and architecture using open-source components have contributed to successfully demonstrating the value of this knowledge management solution for the field force.

This chapter presents the design and architecture of the software. This is followed by a description of the technical solution where we apply the Sunny Aberdeen open-source solution developed by one of the authors, and finally the experiment.

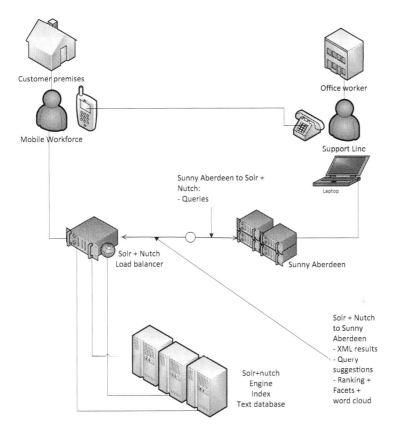

Fig. 8.1 Overall solution design. Solr (Solr is an open-source search engine http://lucene.apache.org/solr) and Nutch (Nutch is an open-source indexer http://nutch.apache.org)

8.2   Software Architecture and Design

The search framework we have developed is a web-based application that provides search and adaptation services for local websites or intranets. Figure 8.2 illustrates the logical components of our adaptive search framework. The framework is built as a web application which can be used as a standalone system or integrated as a web service in other applications as it also provides an XML interface along with an HTML one. The web application interacts with external components which can be easily plugged into the system, i.e. the architecture is made flexible and extensible to other platforms.

8.2.1   The Web Application

The web application has three main engines:

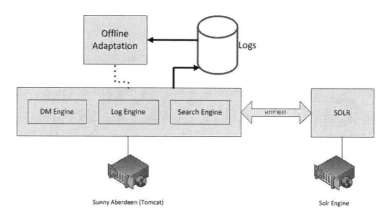

Fig. 8.2 Architecture

- The Search Engine

   The search engine is a wrapper around an open-source search engine that can
   index documents and produce search results for a user query. It is responsible
   for retrieving a ranked list of documents for a user query.

- The Log Engine

   The log engine is responsible for logging all user interactions with the search
   framework. Whenever a new session starts, all the interactions of the user
   within that session are logged. This includes the queries that had been issued
   with their timestamps, the documents that have been retrieved and interactions with the various visual components and their timestamps.

- The Domain Model (DM) Engine

   The domain model engine is responsible for maintaining and adapting a knowledge structure that reflects the document collection and the user community's
   interest. It is consulted to produce query recommendations and to visualise
   these recommendations.

8.2.2   The External Components

Each of the engines within the web application interacts with the corresponding
external resource:

- Apache Solr

   Apache Solr is a popular open-source enterprise search platform from the
   Apache Lucene project. It runs as a standalone web service to index the
   documents on the enterprise or local intranet and provides an interface to

search the index. In addition to that, it has powerful search features such as faceted search and spell checking.

- The Logging Database

    This is implemented as a relational database and hosted on a MySQL service. The logging engine commits all the interactions to this database.

- Offline Adaptation

    This component is responsible for updating the knowledge structure that is maintained by the domain model engine. This component exploits and extends research and implementation from the AutoAdapt project (Lungley and Kruschwitz 2009; Dignum et al. 2010; Kruschwitz et al. 2011; Albakour et al. 2011). The AutoAdapt project has developed a number of methods to learn and adapt domain models from user interactions with a search engine. This is considered the added value in this search framework over a standard one as it automatically updates the domain model from the user community's interests that can aid future users to the right path to finding information in addition to assisting them in browsing and navigation.

8.2.3   The User Interface

Alongside the search results displayed as a standard web search engine results, a number of interactive features are presented to the user. These include a list of query refinements, faceted browsing filters and a tag cloud of query refinements generated using the domain model adapted from previous logs of user interaction. Other interfaces can be instantiated from the framework such as an interactive connected graph of query suggestions.

Interactive information retrieval has received much attention in recent years (e.g. Baeza-Yates et al. 2004). This can also be observed in the major Web search engines that have all become more interactive. A lot of this interactivity comes in the form of query recommendation suggestions. Such suggestions serve different purposes, they allow the user to rephrase the original query, they provide an insight into what the document collection might have to offer, and they illustrate what might be popular information requests. Ultimately they aim at shortening the interaction between user and system and guiding a user to the information that actually satisfies the original information request.

But what suggestions should be recommended to guide the user in the search process? Library classification schemes like the *Universal Decimal Classification*[1] (UDC) have been used for decades and have been demonstrated to be very useful when classifying books. The drawback that these manually encoded classification schemes have is that they lack flexibility. Furthermore, they represent a structured

---

[1] http://www.udcc.org/

view of the world but that view may not be the view that an *online* searcher has. Instead, it has been recognised that there is great potential in mining information from query log files in order to improve a search engine (Silvestri 2010). Given the reluctance of users to provide explicit feedback on the usefulness of results returned for a search query, the automatic extraction of implicit feedback has become the centre of attention of much research (Clark et al. 2011). Queries and clicks are interpreted as 'soft relevance judgments' (Craswell and Szummer 2007) to find out what the user's actual intention is and what the user is really interested in. This knowledge is then used to improve the search engine either by deriving query modification suggestions (the focus of our research) or by using such knowledge to modify the ranking of results to make the search system adaptive to the user or the entire user population.

Whilst query suggestions may seem useful for web search, they can become essential to cut down the time spent on finding documents in enterprise search contexts where users are not just interested in getting *some* matching documents but where they are looking for specific documents, memos, spreadsheets, etc. Enterprise search is very different from web search (Hawking 2010; White 2007; Sherman 2008), and one of the benefits an enterprise search setting offers is access to spam-free document collections as well as implicit feedback from an expert population (very different from the heterogeneous user population searching *the* Web). Our research aims at capturing the implicit feedback from searchers in an enterprise setting to learn query suggestions as part of a continuous learning cycle. The main challenge we face is the dynamic nature of the document collection which requires a constant update of the 'models' that have been learned.

Our particular research is concerned with learning from the interaction logs that we collect from expert searchers who are answering calls in a call centre. The aim is to learn suggestions over time to ultimately shorten the time it takes to answer these calls. We investigate three different learning approaches that turn implicit feedback into models for query recommendation. In this first study we aim to answer two research questions as follows:

- Do the query suggestions presented by any of the models improve over time?
- Does the continuous learning cycle lead to significantly shorter interactions?

Due to the lack of explicit feedback in the experimental setup, we make the simplifying assumption that higher take-up indicates better suggestions (to address the first question).

We report on a longitudinal study in a live industry setting in which we employ three different learning methods to derive query suggestions over time, one using maximum likelihood estimates, one using an ant colony optimisation approach and a third one using association rules. We employ A/B testing to assess these methods and compare them against each other.

8.3 Related Work

Searchers can have difficulties in identifying useful terms that truly represent their information needs when they interact with a search engine (Ruthven 2003). *Query recommendation* addresses this particular issue by explicitly suggesting related queries for a searcher's initial query. It was found to be a promising direction for improving the usability of Web search engines (Joho et al. 2004; White and Ruthven 2006; White et al. 2007). Therefore there has been a lot of interest in the IR community to develop and evaluate query recommendation systems. In particular, many approaches have been proposed to exploit past interactions of the users with the search engine as recorded in the logs to capture 'collective intelligence' for providing query recommendations. Early work investigated the power of association rules applied on queries in the same session to derive related queries (Fonseca et al. 2003). More elaborate techniques that learn from the flow of queries within a session were also proposed. This includes performing random walks on the query-flow graph (Boldi et al. 2008), projections on the query-flow graph (Bordino et al. 2010) or the probabilistic model to predict the next user query introduced in He et al. (2009). Click-through information can also be exploited for query recommendation systems. For example, Mei et al. (2008) developed an algorithm to derive query suggestions from the query-click graph using random walks with hitting time. Ma et al. (2008) used the bipartite graph to derive a query similarity graph with latent semantic indexing.

However, suggestions derived using the previous methods treat the search logs as one batch and do not take the temporal aspect into account. Hence, the recommendations may become out of date and therefore irrelevant. Illustrated in Baraglia et al. (2010) is a query-flow graph model which can become outdated over time. In Broccolo et al. (2010), we can find an incremental version of the 'association rules' and 'query-flow graph' methods where the model can be updated whenever a new query is submitted. In this chapter we will be looking at query suggestion techniques which are adaptive over time.

Developing an efficient and effective search engine for a modern enterprise can have a significant financial value (Feldman and Sherman 2001). Also, as indicated, the characteristics of enterprise search impose a lot of challenges on IR systems and make it different from Web search (Broder and Ciccolo 2004; Hawking 2004).

The search tasks are also different in nature as they are motivated by work problems. This includes approving an employee travel request, responding to a call in a call centre or finding people with technical expertise in a certain topic. Moreover, content in the enterprise comes from different repositories in different formats (emails, Web pages, database records) and do not necessarily cross-reference each other with hyperlinks as in the Web. For example, Fagin et al. (2003) found out that on the IBM corporate intranet, the ratio of strongly connected components (the ones that reach others by following links) accounts for only 10 % of the intranet collection which is much smaller than on the Web. Hence, the popular PageRank (Brin and Page 1998) algorithm for Web search may not be as effective on intranets. For example, Hawking et al. (2004) found out that links on the Web from outside an organisation have little value to add to the quality of the Web site search for that organisation. Mukherjee and Mao (2004) outlined the key ingredients that should be found in an enterprise search system. In addition to various techniques of designing effective ranking models for enterprise search, they suggested that contextualisation and recommendation can be particularly useful for search on the enterprise. In this work, we will take these observations forward and we propose and investigate adaptive domain models for a search engine serving the mobile workforce of a large corporation to improve retrieval in these environments. These models will be able to capture context and assist users in finding the right terminology that should deliver the 'right' answer to their information needs.

8.4 Search Framework to Support the Mobile Workforce

This section presents the context, in which the experiment has been conducted, and then the solution is described in detail and the different algorithms tested are explained.

8.4.1 Problem Description

We start here by describing the context of our solution. We have provided a search engine for a mobile workforce. We have exposed this solution to an operational mobile workforce with fifty desktop-based users supporting about one thousand mobile employees. As discussed earlier, the framework offers adaptive query suggestions to facilitate user experience, which is the major added value in our solution.

The solution we are discussing is designed to be used by top-level support line operators. They have been given a knowledge management tool which lets them access the main technical repository, but also health and safety, and compliance information. This tool also allows them to visualise historical data about specific cases. Mobile workers facing a risk of task failure are required to simply call the expert line for support. Search is paramount in deciding what type of response is offered to the mobile worker in difficulty. Operators will typically offer the following level of support: information (which can be technical, practical, compliance, health and safety, etc.), co-worker support (remotely or in person, typically a colleague close by) or the dispatch of an expert (as and when available). Therefore this solution aims primarily at reducing the number of repeat faults (attempted fixes of faults that fail, and which require new appointments). Of course, cost varies a lot depending on the response level given, and therefore savings can be evaluated in respect to the number of cases that do not fail and the level of response that was required.

8.4.2   System Setup

The search framework we provide is not accessible for mobile workers themselves. This can be seen as a paradox since more powerful mobile devices are becoming available. There are a number of reasons that mobile workers are not being asked to access the system directly: (1) wireless network access can still be poor in many places; (2) mobile devices do not always have a screen that is large enough especially when many documents and maps are involved; and (3) browsing for information might not be the quickest or safest way to carry out a task in jeopardy. The operators are expected to answer most questions without the need to look up the information. Mainly, search queries relate to technical references (e.g. clearance height, fault codes, validation codes) or handling of new equipment (e.g. installation settings), but also jeopardy situations (e.g. hostile environment, equipment incompatibilities, etc.).

Our search solution is therefore required to access a variety of sites which hold technical information, but also health and safety, and compliance. In large corporations such portals or sites exist already, but they may be in a disparate form; they may use recent and rich information management tools (such as SharePoint), or they may be based on basic HTML pages. In any case, they will be attached to a number of processes and safeguards because the information they hold can be business critical or may have legal implications. Replacing and merging such systems is a risky and costly operation. Yet, having one single portal to reach out to all of those systems has obvious benefits in our context.

Deploying an open-source solution and being able to sustain legacy systems has two new benefits: it is low cost and it can be done quickly. We have used Solr,[2] an industry-strength open-source search engine which features rich user interface interaction. It is set up with facets, word cloud, spell checking and configurable ranking for different repositories. As well as aggregating key portals, our knowledge management solution includes a wiki used to keep a record of problems encountered by support line operators. Gradually, this system is growing a knowledge base of difficulties met by mobile workers with the solution applied. The system is inspired by bug tracking software. It is anticipated that the knowledge base hence gathered will be integrated into the search engine; however, it is not the case at this stage. This is due to a possible breach of compliance until the right validation processes can be created.

---

[2] http://lucene.apache.org/solr

8.5   Adaptive Query Suggestions

The search engine has been improved by the addition of a query suggestion engine called Sunny Aberdeen. As discussed earlier, the major added value in our framework is the domain model engine, which derives and adapts query suggestions over time. In particular, we deploy the ant colony optimisation (ACO) approach developed in the AutoAdapt project (Albakour et al. 2011). Here we give a brief description of the ant colony optimisation approach. ACO has been studied extensively as a form of swarm intelligence technique to solve problems in several domains such as scheduling. Here, the ACO analogy is used to first populate and then adapt a directional graph in which the nodes represent queries and the edges represent possible reformulations from one query to another. ACO uses the log of historical sessions as a training data set. A session is a sequence of queries that have been captured under the same browsing session. In an ideal scenario, each session would describe one search attempt, but in practice we expect noise which happens when users perform several distinct search tries within one session (abnormally long sessions have been excluded for the comparison). Automatically identifying the boundaries of sessions is a difficult task (Göker and He 2000; Jansen et al. 2007). One of the reasons is that a session can easily consist of a number of *search goals* and *search missions* (Jones and Klinkner 2008). This is a particular problem in call-centre scenarios. Identifying topically related chains in user query sessions has been studied extensively (Gayo-Avello 2009).

In the ACO analogy the edges in the graph are weighted with the pheromone levels that the ants, in this case, users, leave when they traverse the graph. The user traverses a portion of the graph by using query reformulations identified in a user session (analogous to the ant's journey); the weights of the edges on this route are reinforced (increasing the level of pheromone). Over time all weights (pheromone levels) are reduced by introducing some evaporation factor to reflect unpopularity of the edge if it has not been used by ants. In other words, we reduce the weight of non-traversed edges over time, to penalise incorrect or less relevant query modifications. In addition we expect outdated terms to be effectively removed from the model, i.e. the refinement weight will become so low that the term will never be recommended to the user.

8.5.1   Experiments

The aim of our live experiments is to compare our ACO algorithm with a couple of baseline alternatives, namely, maximum likelihood estimation (MLE) and association rules (AR). In the following, we briefly describe each baseline:

- MLE: MLE is a commonly used baseline approach in statistical natural language processing (Manning and Schütze 1999). This method essentially ranks queries according to how likely they follow up a given query as observed from the past logs.

- AR: Association rules were used as an approach to derive query suggestions by Fonseca et al. (2004). Their approach maps the problem of finding association rules in customer transactions to the problem of finding related queries in a Web search engine. The general intuition is that if distinct queries co-occur in many sessions, then this may be an indication that these queries are related.

The experiment aims to establish which is the most effective algorithm for generating adapting query suggestions and what are the benefits, if any, in terms of speed. The domain model engine responds to the initial query by giving a list of possible words or phrases which are presented as candidates to either replace or augment the initial query. Given the call-centre context, we use a session timeout of 5 min, i.e. a session expires after 5 min of inactivity. To compare the different algorithms, we apply A/B testing (Kohavi et al. 2007).

The domain model engine returns its list from one of the three algorithms randomly. Which algorithm is being used is obviously recorded within the session log. In order to assess the performance of each algorithm, we measure and compare the number of times end users select query suggestions and view between one and three documents returned. We therefore exclude sessions where no documents were viewed or when too many documents were accessed. This is an attempt to make sure that only successful queries are being compared. We do not have a means of knowing whether a search is in fact successful or not from explicit user feedback. Based on the specific context we operate in, we apply the assumption that viewing a small number of documents equates to a successful search. Whilst this is not a perfect match, it is designed to reduce noise to a minimum.

8.5.2 Results and Discussion

The search engine general usage figures show us that it processes on average just over 600 queries per week. The average session length is 1.92 queries (with a 1.83 standard deviation). This is shorter than an average Web query, e.g. (Silverstein et al. 1999). Sessions of one query only account for 62 % overall, and sessions of two or three queries represent 27 % of all queries; longer sessions (four queries or more) are represented by the remaining 11 %. Finally, it takes an overall average of 62 s to access documents from initial query to final document viewing. These figures are only shown here to give an idea of the scale of the application; they are not relevant to the experiment in itself.

The experiment is divided into two phases: a training phase where algorithms are gathering data and a second post-training phase where algorithms have been trained and where we expect better suggestions. We then compare usage between the two phases. We expect to see an improvement in the quality of the results when users exploit query suggestions.

Fig. 8.3 A/B testing for query success ratio

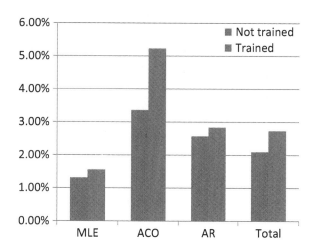

In Fig. 8.3 all algorithms start untrained and are trained on 30/07/2012. We consider the unpaired t-test analysis to be positive if its result is below 0.05. The results are ACO = 0.019, AR = 0.209 and MLE = 0.211.

We run an unpaired t-test analysis to measure whether or not there is a significant difference in usage. Figure 8.3 plots usage for each algorithm. The period before 30/07/12 is the training phase; the period thereafter shows results after training. The t-test analysis is only positive for the ACO algorithm. The MLE model does not indicate a significant usage increase and it remains quite low overall. The AR model is very much irregular in its usage perhaps indicating that suggestions vary in quality.

8.5.3   Time Comparison

The second part of the comparison aims to understand if using query suggestion can help save time. The query suggestion concept reduces the amount of typing and may also guide users during the course of their search sessions. We have measured successful queries in an attempt to establish whether suggestions lead to shorter sessions.

We measured the time length of sessions by subtracting the timestamp of the initial query from the timestamp of the last viewed document. We excluded sessions where more than three documents were found and session that spans over 5 min because this is a typical indication that the same session is being used for separate queries. Figure 8.4 shows three average session times. The first one is the time spent for a single, successful query which is 27 s. A query refined manually takes on average 81 s. A query which contains both manual refinements and suggestions takes on average 85 s. Finally, a query refined using suggestions takes 79 s. Time taken to refine queries is not improved significantly.

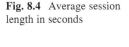

Fig. 8.4 Average session length in seconds

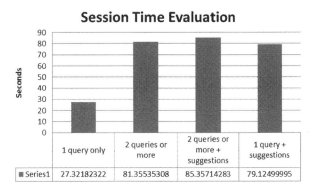

Session Time Evaluation

	1 query only	2 queries or more	2 queries or more + suggestions	1 query + suggestions
▦ Series1	27.32182322	81.35535308	85.35714283	79.12499995

8.6   Conclusions

This setup demonstrates a design and high-level architecture for a practical mobility and knowledge management solution. The originality is to use an intermediate party to validate information going to field workers. This means that the tool focuses on empowering desktop-based workers rather than mobile workers. Desktop-based workers are more likely to be open to powerful interfaces, and they can improve their information-seeking skills with the provided search.

The benefits, although hard to quantify, can be listed as follows: access to technical documentation for unexpected difficulties leads to less repeat faults; the ability to monitor and enable collaborative work centrally leads to more effective use of resources. Access to all relevant aspects of unexpected tasks (health and safety, regulatory and technical) leads to compliant practices. Of course the search solution brings a significant contribution to the provision of a cost-effective functionality.

We weren't able to measure the effective reduction in repeat faults because of high variability in this data set. Usage gives an indication of whether the search is deemed useful or not. There log-based evidence of sustained usage and indexing new content (to make new material available as soon as possible) has been a recurrent issue which is another indicator that search plays an important part.

We also introduce a learning element which brings huge benefits in terms of reducing maintenance costs. Enterprise search solutions are notorious for the level of manual configuration they require. Automated learning simplifies software maintenance and therefore reduces overheads. We identify a stronger learning model called ACO.

The relatively low usage of query suggestion (one in forty queries) might indicate that there are not enough suggestions made to represent the complexity of all queries at this stage. The ACO model is also more flexible in its ability to learn and discard query suggestions making it more robust and sustainable.

General usage reflects an important challenge in enterprise search query suggestion, in many cases data can remain sparse, and even weeks of training do not lead to an optimum trained set. Further work is required to understand if and when the algorithm peaks with regard to usage. We need to establish a ratio between usage and quality against which we would be able to compare algorithms. Other solutions might be also be used to supplement training and increase usage more significantly.

For various reasons there is a growing interest in query suggestions. People are expecting quality to equal mainstream Internet search engines that are able to process large quantities of information. This work demonstrates valuable ways to progress at the enterprise search level. Ultimately, we believe that this will also lead towards a very valuable 'queryless' search with a high degree of personalisation. This potential could play an important role in enterprise search and mobility.

References

Albakour M-D, Kruschwitz U, Nanas N, Song D, Fasli M, de Roeck A (2011) Exploring ant colony optimisation for adaptive interactive search. In: Proceedings of the third international conference on the theory of information retrieval (ICTIR). Lecture notes in computer science. Springer, Bertinoro, pp 213–224

Baeza-Yates R, Hurtado C, Mendoza M (2004) Query recommendation using query logs in search engines. In: Proceedings of ClusWeb'04. Springer, pp 588–596

Baraglia R, Nardini FM, Castillo C, Perego R, Donato D, Silvestri F (2010) The effects of time on query flow graph-based models for query suggestion. In: RIAO. pp 182–189

Boldi P, Bonchi F, Castillo C, Donato D, Gionis A, Vigna S (2008) The query-flow graph: model and applications. In: Proceeding of CIKM'08. ACM, New York, NY, pp 609–618

Bordino I, Castillo C, Donato D, Gionis A (2010) Query similarity by projecting the query-flow graph. In: Proceedings of SIGIR'10, Geneva. pp 515–522

Brin S, Page L (1998) The anatomy of a large-scale hypertextual Web Search Engine. In: Proceedings of the seventh international World Wide Web Conference (WWW7), Brisbane. pp 107–117

Broccolo D, Frieder O, Nardini FM, Perego R, Silvestri F (2010) Incremental algorithms for effective and efficient query recommendation. In: Proceedings of the 17th international symposium string processing and information retrieval SPIRE'10. pp 13–24

Broder AZ, Ciccolo AC (2004) Towards the next generation of enterprise search technology. IBM Syst J 43(3):451–454

Clark R, Kim Y, Kruschwitz U, Song D, Albakour MD, Dignum S, Cervino Beresi U, Fasli M, de Roeck A (2011) Automatically structuring domain knowledge from text: an overview of current research. Inform Process Manag 48:552–568. doi:10.1016/j.ipm.2011.07.002

Craswell N, Szummer M (2007) Random walks on the click graph. In: Proceedings of the 30 th annual international ACM SIGIR conference on research and development in information retrieval, Amsterdam. pp 239–246

Dignum S, Kruschwitz U, Fasli M, Kim Y, Song D, Cervino U, de Roeck A (2010) Incorporating seasonality into search suggestions derived from intranet query logs. In: Proceedings of the IEEE/WIC/ACM international conferences on Web intelligence (WI'10), Toronto, pp 425–430

Fagin R, Kumar R, McCurley KS, Novak J, Sivakumar D, Tomlin JA, Williamson DP (2003) Searching the workplace web. In: WWW. pp 366–375

Feldman S, Sherman C (2001) The high cost of not finding information. Technical Report 29127. IDC

Fonseca BM, Golgher PB, de Moura ES, Ziviani N (2003) Using association rules to discover search engines related queries. In: Proceedings of the first Latin American Web Congress. pp 66–71

Fonseca BM, Golgher PB, de Moura ES, Pôssas B, Ziviani N (2004) Discovering search engine related queries using association rules. J Web Eng 2(4):215–227

Gayo-Avello D (2009) A survey on session detection methods in query logs and a proposal for future evaluation. Inform Sci 179:1822–1843

Göker A, He D (2000) Analysing web search logs to determine session boundaries for user-oriented learning. In: AH '00: Proceedings of the international conference on adaptive hypermedia and adaptive Web-based systems. Springer, Berlin, pp 319–322.

Hawking D (2004) Challenges in enterprise search. In: ADC. pp 15–24

Hawking D (2010) Enterprise search. In: Baeza-Yates R, Ribeiro-Neto B (eds) Modern information retrieval, 2nd edn. Addison-Wesley, Harlow, pp 645–686

Hawking D, Crimmins F, Craswell N, Upstill T (2004) How valuable is external link evidence when searching enterprise webs? In: ADC. pp 77–84

He Q, Jiang D, Liao Z, Hoi SCH, Chang K, Lim EP, Li H (2009) Web query recommendation via sequential query prediction. In: Proceedings of the 2009 I.E. international conference on data engineering, ICDE '09. IEEE Computer Society, Washington, DC, pp 1443–1454

Jansen BJ, Spink A, Blakely C, Koshman S (2007) Defining a session on Web search engines. J Am Soc Inform Sci Technol 58(6):862–871

Joho H, Sanderson M, Beaulieu M (2004) A study of user interaction with a concept-based interactive query expansion support tool. In: ECIR. pp 42–56

Jones R, Klinkner KL (2008) Beyond the session timeout: automatic hierarchical segmentation of search topics in query logs. In: Proceeding of the 17th ACM conference on information and knowledge management (CIKM'08). pp 699–708

Kohavi R, Henne RM, Sommerfield D (2007) Practical guide to controlled experiments on the web: listen to your customers not to the hippo. In: KDD. pp 959–967

Kruschwitz U, Albakour MD, Niu J, Leveling J, Nanas N, Kim Y, Song D, Fasli M, de Roeck A (2011) Moving towards adaptive search in digital libraries. In: Advanced language technologies for digital libraries. Lecture notes in computer science, vol 6699. Springer, Heidelberg, pp 41–60

Lungley D, Kruschwitz U (2009) Automatically maintained domain knowledge: initial findings. In: Proceedings of the 31st European conference on information retrieval (ECIR'09), Toulouse. pp 739–743

Ma H, Yang H, King I, Lyu MR (2008) Learning latent semantic relations from clickthrough data for query suggestion. In: Proceedings of the 17th ACM conference on information and knowledge management, CIKM '08. pp 709–718

Manning CD, Schütze H (1999) Foundations of statistical natural language processing. MIT, Cambridge, MA

Mei Q, Zhou D, Church K (2008) Query suggestion using hitting time. In: Proceedings of the 17th ACM conference on information and knowledge management, CIKM '08. pp 469–478

Mukherjee R, Mao J (2004) Enterprise search: tough stuff. Queue 2(2):36–46

Ruthven I (2003) Re-examining the potential effectiveness of interactive query expansion. In: Proceedings of the 26th annual international ACM SIGIR conference on research and development in information retrieval, Toronto, Canada. pp 213–220

Sherman C (2008) Why enterprise search will never be Google-y". In: Enterprise search sourcebook. pp 12–13

Silverstein C, Henzinger MR, Marais H, Moricz M (1999) Analysis of a very large web search engine query log. SIGIR Forum 33(1):6–12

Silvestri F (2010) Mining query logs: turning search usage data into knowledge. Found Trend Inform Retriev 4:1–174

White M (2007) Making search work: implementing Web, intranet and enterprise search. Facet, London

White RW, Ruthven I (2006) A study of interface support mechanisms for interactive information retrieval. J Am Soc Inform Sci Technol 57(7):933–948

White RW, Bilenko M, Cucerzan S (2007) Studying the use of popular destinations to enhance Web search interaction. In: Proceedings of the 30th annual international ACM SIGIR conference on research and development in information retrieval. Amsterdam, pp 159–166

Chapter 9
Application of AI Methods to Practical GPON FTTH Network Design and Planning

Kin Fai (Danny) Poon, Anis Ouali, Andrej Chu, and Riaz Ahmad

Abstract The optimal design of telecommunication network infrastructure demands consideration of many complex factors such as type, number and position of components and cable paths. The difficulty of producing a consistent and cost-effective solution increases with network size and complexity. A network planning optimisation tool developed at EBTIC (Etisalat British Telecom Innovation Centre) can generate different network topologies and evaluate them to arrive rapidly at an optimal or near-optimal solution. It has been trialled in Greenfield areas for FTTH (Fibre To The Home) deployment. This case study reviews the operation of EBTIC optimisation tool, how the design problem has been formulated and summarises the key steps in the practical application of the algorithms. In addition, the business benefits provided by the tool in the working environment are provided.

9.1 Introduction

Internet traffic has experienced a significant growth due to bandwidth hungry applications such as online gaming, Internet TV, video streaming and real-time social networking. The trend will only continue, demanding from the communication infrastructure higher speed, greater capacity and longer reach. In the past, tremendous progress was made to improve the communication bandwidth in optical networks. While the advancement in optical communication technology has made a significant progress, network design of physical layer still plays an important role in the effectiveness of an entire optical system. Not only does a poor network design incur a higher deployment and management cost, but it can also underutilise the full potential of a fibre-based system.

In addition, the increasing competition drives the need to provide services more quickly in a cost-effective manner. The costs associated with the design and build of

K.F. (Danny) Poon (✉) • A. Ouali • A. Chu • R. Ahmad
Etisalat-BT Innovation Centre (EBTIC), Khalifa University, Abu Dhabi, UAE

G. Owusu et al. (eds.), *Transforming Field and Service Operations*,
DOI 10.1007/978-3-642-44970-3_9, © Springer-Verlag Berlin Heidelberg 2013

new networks are high, and upgrading existing networks to provide the required services can be very expensive. Furthermore, the disruption incurred in building a new network or upgrading an existing one should be minimised as much as possible. A well-planned network is also the key to minimise investment, improve the average profit per connected user and speed up the return on investment.

According to the market research report by TechNavio (2011–2015), '*the Global Passive Optical Network Equipment market to grow at a CAGR of 29.28 percent over the period 2011–2015*', therefore, the potential for cost savings through improvements in network design, build and operation can be substantial. For example, an implementation of Fibre To The Home (FTTH) network in a new big area costs hundreds of millions of dollars. If the deployment cost can be reduced even by only 10 % (e.g. by minimising digging up the road for laying cables and ducts and optimising the number of optical splitters required), the saving can still be significant.

Due to the continued development of new housing estates, there is always a need to reduce the cost of provision of customer service on FTTH access networks. Typically, a GIS (Geographical Information System) planning software is used to design the network manually. The input to the GIS system is a base map of a particular area detailing the locations of premises and road networks. Depending on the size of the given area, the planner may require to segment the entire area into individual catchment areas. The size of each catchment area is based on the number of required connections and the capacities of different network components. For example, if each Fibre Distribution Hub (FDH) can accommodate ten 1:32 optical splitters, a maximum of 320 connections can be provided for a given area.

This manual planning process relies heavily on the planners' skill, expertise and judgement in the interpretation of planning policies and rules. This situation can possibly result in creating less optimised designs, which can incur a higher deployment cost. Initial development work on optimisation models for FTTH Greenfield networks had proved that savings could be achieved if the optimisation functionality was applied to network design. During the development phase, a decision of how the functionality could be deployed was made. There were two possible options: (a) to provide a standalone application which allows planners to design FTTH networks using the optimisation system or (b) to integrate the optimisation module within the existing GIS network planning system. Since a large amount of money on the existing planning system has been invested and all the planners have been trained to use the system, the latter option was chosen (i.e. the optimisation module was deployed as a web service).

9.2 Related Work

Due to the drastic increase in FTTH network deployment, many telecom companies and academia are striving for solutions to minimise the planning cost. Therefore, many network optimisation approaches have been used and can be categorised into two main categories: classical approach based on the mathematical programming and meta-heuristics such as Genetic Algorithm (GA), Simulated Annealing (SA) and Ant Colony Optimisation (ACO). Both approaches have pros and cons.

For a small/medium sized network without too many constraints, mathematical programming (i.e. Mixed Integer Linear Programming (MILP) or Binary Integer Programming (BIP)) usually outperforms meta-heuristics in terms of the computational performance to obtain an optimum or near-optimum solution. However, when the number of the variables is very large due to the size and the optimisation complexity of the given network, meta-heuristics (with a local search) usually perform better (Cambazard et al. 2011). It is due to the fact that the time required to traverse the search tree in the MILP case can be very significant. Another advantage of employing meta-heuristics is the adaptability to new requirements. Since most of the meta-heuristics can be treated as a black box without knowing the landscape of the problem, new requirements can be added relatively easy by modifying the objective function. In the following, we describe some related work with regard to both approaches.

For the mathematical optimisation, the authors in Li and Shen (2009) formulate the Passive Optical Network (PON) planning problem using a non-linear mathematical model to minimise the overall deployment cost. A problem of locating or allocating splitters for the fibre access network using MILP approach with reformulation–linearisation technique to provide a tighter representation can be found in Kim et al. (2011). A combination of the clustering algorithm based on the Single-Linkage Algorithm (SLA) (Penrose 1995) to group the Optical Network Units (ONUs) and the Linear Optimisation technique to select or position the splitter/arrayed waveguide grating (AWG) for Wavelength Division Multiplexed (WDM) PON network is detailed in Jaumard and Chowdhury (2012). In addition, the authors state that the two-phase approach demonstrates the effectiveness of the proposed solution. In Poon and Ouali (2011), a MILP approach is employed to optimise the FTTH access networks with the consideration of future growth. Different planning scenarios were considered to investigate the impact of the future demand and the associated planning cost.

Regarding the meta-heuristic approach for the FTTH network planning, in Lv and Chen (2009), the authors developed an optimisation solution to perform multi-hierarchy PON planning. In their case, upper and lower Optical Branching Devices (OBDs) were introduced. The upper OBDs were used to connect between Optical Line Terminal (OLT) and lower OBDs while lower OBDs connected between ONUs and upper OBDs. The locations of OBDs were calculated based on the Max–Min Distance Cluster (MMDC) algorithm which can also be found in Lv and Chen (2009). A Genetic Algorithm (GA) was then applied to perform the routing optimization process. A city block map was first converted into a grid map. Each horizontal or vertical line intersected at least one node in the grid. From each node, four possible directions could be established to route to another node. The GA was employed to find the optimum routings to traverse all the selected nodes.

In Steve (2008), the author considered four access network architectures/technologies: Digital Subscriber Line (xDSL) from the central office, Fibre To The

Premises (FTTP), Fibre To The Node (FTTN) and Fibre To The Micro-Node (FTTn). Different technologies were combined to reduce the copper distance by installing DSL access multiplexers in outdoor cabinets and field micro-nodes which were closer to the subscribers. In addition, different classes of services and different subscribers per class per point of demand were considered. A heuristic algorithm based on tabu search was proposed to find good feasible solutions within a reasonable amount of computational time.

The authors in Li and Shen (2009) focused on minimising the planning cost of a Greenfield long-reach PON deployment. They proposed a heuristic named Recursive Association and Relocation Algorithm. The basic idea was to randomly generate a solution, to recursively improve by an ONU-Splitter assignment process and to determine the splitters with minimum cost. The convergence of the solution was controlled by a simulated annealing (SA)-like algorithm presented in their paper.

The authors of Poon et al. (2006) applied genetic algorithms to optimise the cost of FTTH planning. Firstly, they proposed a heuristic method to form clusters of PONs. Afterwards, they identified the splitter location of each PON. Finally, they used a genetic algorithm to identify the locations of fibre distribution points to connect ONUs and splitters.

In Lakic and Hajduczenia (2007), the authors designed the FTTH networks through three steps. The first step was related to obstacle avoidance where some paths could not be used to lay fibres. In the second part, they clustered the customer premises based on the K-means clustering algorithm. Finally, they applied genetic algorithms to optimise the fibre path. The splitter locations were given prior to running the algorithms.

9.3 GPON/FTTH Network Structure

There are many FTTH architectures in current deployment around the world today. Some of them are based on point-to-point, point-to-multipoint, multi-level split or some form of hybrid active network. Although the detailed operating characteristics of these networks are different, the architecture is likely to have many design features and aspects in common. For example, networks based on GPON (Gigabit Passive Optical Network) technology will typically have a tree structure with a head-end node, optical splitters (either centralized or distributed throughout the network) and Optical Network Units (ONUs) situated at the customer premises. These networks are installed using various techniques such as using standard or blown fibre/cable, duct distribution schemes utilising duct tees, individual duct or even no duct with direct buried cables and various types of joint/splitter enclosures. Each scheme will have its own set of planning requirements which must be considered and satisfied.

A particular GPON architecture considered in this case study is a centralised network structure. Advantages of deploying this kind of structure (Mazzali

Fig. 9.1 A typical GPON
fibre distribution network

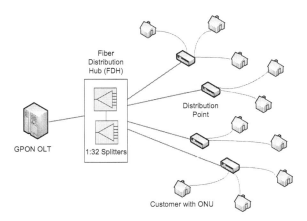

et al. 2005) include single point of maintenance in the network, lower splitter cost-to-ports ratio and less insertion loss due to the absence of concatenating splitters. A typical GPON distribution network is illustrated in Fig. 9.1. It contains Optical Line Termination (OLT) equipment located in the Point of Presence (POP), optical splitters in Fibre Distribution Hubs (FDHs) and a number of ONUs connecting customers to the PON network. As illustrated in Fig. 9.1, each PON provides a maximum of 32 connections. Customers are connected to the splitter through Distribution Points (DPs). DPs are situated relatively close to customer premises and they provide convenient access points for the customer distribution cables.

9.4 Optimisation Objectives and Network Model

The FTTH design problem that we tackle can be described as a kind of a multi-layered hierarchical location and assignment problem. Customer premises are considered as the lower layer while Distribution Point and Fibre Distribution Hub locations are considered as middle and upper layer, respectively. An input network file of the optimisation module can be configured as either a tree or a mesh network. It then determines the best locations of network elements and identifies geographically advantageous tree structure subnetworks to aggregate cables from customers to a POP via DPs and FDHs. A solution for the considered problem should minimise the deployment cost of an FTTH network by optimising the following objectives:

- Number of PONs required by the total demand of customer connections
- Placement of DPs
- Placement of FDHs
- Clustering of customers to DPs
- Clustering of DPs to FDHs
- Required cable sizes

- Routing information of connecting customers to DPs and DPs to FDHs

  Subject to:

- Maximum cabling distances among different network elements
- Capacity limitations of the DPs and FDHs

For each area to be planned, we assume that the following input information is available:

- M—set of customer premises, with a cardinality of m
- N—set of possible cable Distribution Points (DPs), with a cardinality of n
- O—set of possible Fibre Distribution Hubs (FDHs), with a cardinality of o
- A—distance matrix between the customer premises and possible DPs, sized $m \times n$
- B—distance matrix between the possible DPs and possible FDHs, sized $n \times o$
- C—distance vector between the possible FDHs and the POP
- *MaxConnectionsPerDP*—maximum number of tenants that can be connected to a single DP
- *MaxDPsPerFDH*—maximum number of DPs that can be connected to a single FDH
- *MaxDistanceDP*—the constraint on maximum distance of the customer from the assigned DP
- *MaxDistanceFDH*—the constraint on maximum distance of the DP from the assigned FDH node
- *DPunitCost*—the price of a single DP unit
- *FDHunitCost*—the price of a single FDH unit
- *1FcableCost*—the price of 1 m of a 1-fibre cable, used to connect a customer to a DP
- *24FcableCost*—the price of 1 m of a 24-fibre cable, used to connect a DP to a FDH
- *480FcableCost*—the price of 1 m of a 480-fibre cable, used to connect FDH to the Point of Presence (POP)

From the topology point of view and for the purpose of network design, there is additional input information required:

- Location of a POP
- Locations of the customer premises
- Possible locations of the DPs and FDHs
- A civil layer network that specifies the connectivity among the possible locations of the different network elements

The following describes the model in terms of variable and parameters, optimisation criteria and planning constraints.

Variables and Parameters: In the present model, links from customer plots to DPs and from DPs to FDHs are modelled using Boolean variables. Thus, we have two variable matrices *PlotToDPlink* and *DPtoFDHlink*.

PlotToDPlink is defined in $m \times n$ such that

$$PlotToDPlink[i][j] = 1, i \in M, j \in N \text{ iff } i \text{ is assigned } j \quad (9.1)$$

Similarly, *DPtoFDHlink* is defined in $n \times o$ such that

$$DPtoFDHlink[j][k] = 1, j \in N, k \in O \text{ iff } j \text{ is assigned to } k \quad (9.2)$$

In addition, two Boolean variable arrays *UsedDP* and *UsedFDH* (of n and o elements) are introduced to denote whether the corresponding DP and FDH are being used.

Based on the network connectivity and the locations of network nodes, the shortest path between any pair of nodes is computed and stored in *CivilDistance* matrix.

Planning Constraints: Most of the model constraints derive from the planning rules that are commonly used in practice. For example, each plot must be connected to exactly one DP and such a connection must satisfy the distance requirement between a plot and a DP. Similarly, each DP must be connected to exactly one FDH without exceeding the maximum allowable distance.

Other constraints derive from the characteristics of the network equipment. They consist of distance constraints between customers and DPs:

$$PlotToDPlink[i][j] \times A(i,j) \leq MaxDistanceDP, i \in M, j \in N \quad (9.3)$$

and between DPs and FDHs:

$$DPtoFDHlink[j][k] \times B(j,k) \leq MaxDistanceFDH, j \in N, k \in O \quad (9.4)$$

Also, network equipment capacity constraints should be satisfied for DPs:

$$\sum_{i \in M} PlotToDPlink[i][j] \leq MaxDistanceDP, j \in N \quad (9.5)$$

and FDHs:

$$\sum_{j \in N} DPToFDHLink[j][k] \leq MaxDPsPerFDH, k \in O \quad (9.6)$$

In an attempt to provide some spare capacity for future provisioning, both *MaxTenanciesPerDP* and *MaxDPsPerFDH* are set to values that are less than the maximum physical capacity of the corresponding splitters.

Optimisation Criteria: The objective of the present model as shown in Eq. (9.7) is to minimise the global cost of planning the cable layer of a network:

$$GlobalCost = DP\cos t + FDH\cos t + CableCost \quad (9.7)$$

where

$$DPcost = DPunitCost \times \sum_{j \in N} UsedDP[j] \qquad (9.8)$$

$$FDHcost = FDHunitCost \times \sum_{k \in O} UsedFDH[k] \qquad (9.9)$$

and

$$Cable\,\cos t =$$
$$1FcableCost \times \sum_{i \in M} \sum_{j \in N} PlotToDPLink[i][j] \times A(i,j) + 24FcableCost \times$$
$$\sum_{j \in N} \sum_{k \in O} DPToFDHLink[j][k] \times B(j,k) + 480FcableCost \times \sum_{k \in O} C(k)$$

$$(9.10)$$

CableCost consists of the cost of fibre cables from plots to DPs, from DPs to FDHs and from FDHs to the POP. Some other fixed costs may be included but they have no impact on the solution. For instance, the termination cost which includes cost of a drop terminal for each customer plot remains the same as long as the number of plots does not change.

9.5 Network Optimisation Module

To automate the process of the FTTH network design, an ACO (Ant Colony Optimisation)-based algorithm has been employed. Given a set of locations of customer premises, a set of potential locations of cable Distribution Points (DPs) and Fibre Distribution Hub (FDHs), the algorithm is applied to identify the optimal or near optimal locations of DPs and FDHs and the routings of cables from the customer premises to DPs, from the DPs to FDHs and finally from the FDHs to the given location of an exchange. During the optimisation process, the algorithm takes into account all the practical planning rules (e.g. the number of allowable connections for each Distribution Point, capacity of each Splitter Node, maximum distance between different network components, etc.).

The design principle of the ACO was first published in Dorigo and Stützle (2004). It is inspired by the natural behaviour of ants which communicate together using pheromone trails in order to find food paths. The stronger the trail is, the more it attracts other ants. Building a solution is usually realised by constructing a subgraph of the input network. This subgraph needs to meet certain conditions and properties, defined by the problem and the feasible solution. The edges for the solution are usually chosen randomly, with probabilities corresponding proportionally to the pheromone values associated with them.

During the iteration, each ant needs to build a solution, which is evaluated with regard to the objective function. After solution evaluation, the pheromone trails are updated by laying more pheromone on the edges that have been used in the solution found. The amount of pheromone to be laid depends on the quality of the solution—

the better the solution is, the more the ant tries to attract the other ants to it. Standard ACO algorithm uses Eq. (9.11) for calculating the probabilities every time the ant needs to decide the next movement:

$$p_{ij} \leftarrow \frac{\tau_{ij}^{\alpha} \eta_{ij}^{\beta}}{\sum\limits_{k \in N} \tau_{ik}^{\alpha} \eta_{ik}^{\beta}} \qquad (9.11)$$

This equation computes the probability that an ant will move from a node i to a node j based on the heuristic (η_{ij}) and pheromone information (τ_{ij}) associated with the edge (i, j). N_i denotes the neighbourhood of the node i. Heuristic information does not change during the execution of the algorithm and represents the static data, while pheromone information evolves dynamically as it is updated by the ants.

The parameters α and β are used to adjust the weight of the heuristic and pheromone information. This way, it is possible to set the influence of the particular information on the overall probability values. Experiments in Chu (2009) have shown that, for better performance of the algorithm, the pheromone information is most important, while the heuristic information is not that crucial.

In addition, removing the heuristic information from the probabilities, computation has positive effect on the performance. Computing the probabilities is the most frequent operation in the entire algorithm. In order to improve the computational performance of the algorithm, a simplified Eq. (9.12) is used instead:

$$p_{ij} \leftarrow \frac{\tau_{ij}}{\sum\limits_{k \in N} \tau_{ik}} \qquad (9.12)$$

In the initial iterations of the algorithm, the probabilities for each assignment combination are the same, and therefore, it is relatively difficult for the algorithm to find a feasible solution as it cannot use previously acquired information (pheromone paths). In such situation, it might be very helpful to soften the constraints either by raising the upper bound to allow more DPs to connect to a single FDH or by lowering the lower bound to have more customers to be connected to a single DP. A solution obtained under such circumstance is unlikely to be feasible for the given problem, but can help the future iterations to find a feasible solution more quickly. In fact, the pheromone laid on the edges associated with this solution will help to bias probabilities (and consequently assignment decisions) towards the space with feasible solutions. After the given number of iterations has elapsed, the algorithm tightens the bound (or bounds) and the whole process is repeated again several times. Eventually the bounds are set as for the original problem, thus satisfying all the constraints.

9.5.1   Building the Solution

Considering the input as defined in the previous network model, N_i denotes a set of possible DP locations that are reachable by customer i (i.e. the distance between the customer and DP is not exceeding the given constraint *MaxDistanceDP*) and O_j as a set of possible FDH locations reachable (according to the constraint *MaxDistanceFDH*) from DP location j.

The process of building the solution starts from the lowest layer, i.e. the customer premises. Each iteration begins by selecting a random customer $i \in M$. The probabilities of assigning i to j for every $j \in N_i$ are computed using Eq. (9.12). Given the probabilities, a random pick is used to choose the DP to which the customer will be assigned. The whole process is repeated for every customer.

After all customers are assigned to DPs, the same process is repeated on the upper level by assigning selected DPs to FDHs. We can slightly modify Eq. (9.12) to be used for upper level of the input. Let us define $\overline{N}$ as a set of all DPs which have been assigned in the previous process to a customer at least once. Equation 13 is then used to compute the probabilities and perform a random pick for assignment of particular DPs to the FDHs according to the calculated probabilities:

$$\forall j \in \overline{N}, \forall k \in O_j : p_{jk} \leftarrow \frac{\tau_{jk}}{\sum_{l \in O_j} \tau_{jl}} \tag{9.13}$$

9.5.2   Updating the Pheromone Information

After the whole process is finished, the objective value can be computed based on the costs of network components. According to the evaluation of the objective value, the pheromone trails associated with the edges in the graph are updated based on Eq. (9.14), where *TotalCost* is the objective value of the solution found. Thus, the quality of the solution affects the value of the pheromone trails—the better the solution is, the higher will be the updated pheromone value:

$$\forall \, edge \in solution : \tau_{edge} \leftarrow \tau_{edge} + \frac{1}{TotalCosts} \tag{9.14}$$

A concept of pheromone evaporation which is borrowed from the natural behaviour of ant colony is also adopted. The idea is that the pheromone evaporates in the nature under the influence of environment such as temperature and weather. A small amount of pheromone associated with the edge of the graph diminishes in every iteration. Therefore, edges that are unlikely to be included in the best solution are gradually becoming less and less attractive to be selected when computing the probabilities for the next iteration. The concept of pheromone evaporation is implemented easily by adopting Eq. (9.15).

$$\forall\, edge \in solution: \quad \tau_{edge} \leftarrow \tau_{edge} * \rho, \ 0 < \rho < 1 \qquad (9.15)$$

9.5.3 Post-Optimisation

As shown in (Dorigo and Stützle 2004; Chu 2009), post-optimisation plays an important role in ACO method applications. After the solution is built by an ant, it can be inspected for the possible improvements. There are several ways for accomplishing this:

- Introducing a new DP into the solution if possible with regard to the capacity constraints—this adds extra costs to the objective function for another DP used, but can save the cable costs. Therefore, it is needed to investigate whether introducing a particular DP is advantageous.
- Reallocating the customer to a different DP—if there is a close DP which has a free capacity and the distance is shorter than the one from the currently assigned DP, we can reallocate the customer to that selected DP.
- Exchanging of customers—if there is a pair of customers which can exchange their DP assignments (with the consideration of the distance constraint) and this exchange would help to lower the length of cable used.
- Exchanging of DPs—if it is possible, we can reassign all customers of a single DP to a different DP. The cost of DPs would remain the same but we might save the costs of the cables.

All these post-optimisation tasks are also performed one level up, between DPs and FDHs. In this case, the post-optimisation steps are performed on much smaller sets and therefore the process is significantly faster. These heuristics explore the search space around the current solution in order to find a solution with a better value of the objective function. In our implementation, only the search space which generates feasible solutions is considered (e.g. assigning a customer to another reachable DP).

9.6 Software Deployment Approach

The automated FTTH design functionality is provided through the web service technology. The existing GIS planning system looks exactly the same as before with a few more buttons to access the value-added features. This approach has many benefits, for example:

- It provides a cost-effective solution to enhance the existing system independently of system platform and programming languages.
- It eliminates the needs to retrain the planners to adopt a new planning system, which increases the time and the cost of operation.

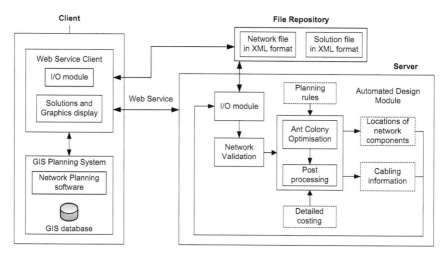

Fig. 9.2 FTTH design web service architecture

- More processing power can be provided by the server to execute the computationally intensive optimisation independently of the resource available to the user's local machine.
- Any update within the optimisation module doesn't require reinstallation of the client application. A new web service can be redeployed in the server.

Figure 9.2 provides a high-level overview of FTTH design module and its interactions with other components in the system.

The client application of the web service, which resides within the GIS planning system, represents the interface between the related planning application and the design module. It provides the following main functions:

- Allowing the user to define the network area which needs to be planned and to change the parameters of the optimisation. Examples of parameters include the capacity of network equipment, maximum execution time and maximum allowable distance.
- Exporting the information of the network selected by the user in a XML document following a given schema.
- Invoking the web service—while waiting for the results, the user may check the current status of the optimisation process or may decide to stop the execution.
- Importing the result which is in XML format back to the GIS planning system.
- Providing the necessary features for the user to display and interact with the results, for example, showing the assignments among network equipment and the corresponding cabling paths.

The network design module, which is embedded in the server side, consists of three main sub-modules: 'I/O module', 'Validation module' and 'Optimisation module'. The 'I/O module' fetches a given input file and executes an XML parser

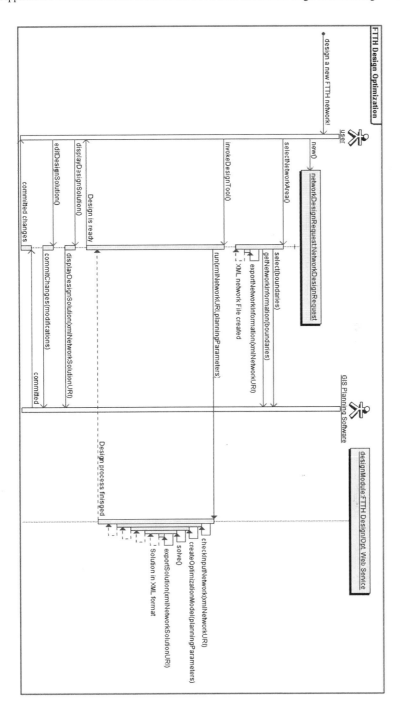

Fig. 9.3 Simplified sequence diagram of the FTTH design web service

to extract the necessary information to create a specified network model. It is also applied to generate the XML document of the design solution. The 'Validation module' checks the connectivity of the given network and the feasibility of the optimised solution. If the 'Validation module' detects any error, the design process will be terminated. The 'Optimization module' is responsible of generating the network design automatically while taking account of network information, planning parameters, design rules, cost and different types of network components. The result, which comprises the optimised locations and assignments of network components, will be exported to the client application to display the solution within the GIS network planning system.

Figure 9.3 depicts a simplified UML sequence diagram describing the interactions between the components of the system in order to achieve the design of the selected area.

The user and the existing planning software are considered as actors. The *networkDesignRequest* instance represents the client side of the web service, while the instance of *designModule* resides in the server side. The sequence diagram shows three main steps of executing the FTTH design web service.

The first step deals with the preparation of network information. The user selects a network area which will be planned. The GIS planning system analyses the selected area and invokes the *networkDesignRequest* object in order to export the information in XML format. The second step is to execute the web service upon request. First, the user sends a request of the *networkDesignRequest* object to proceed with the design of the network. A web service request then invokes the *designModule* object to run all the required tasks. This includes verifying the input network, creating a network model and optimising the network design by minimising the overall cost. Once the optimisation process is finished, a solution file will be exported as an XML file. The web service response is then sent to the caller, *networkDesignRequest* object. The final phase is related to the visualisation and the modification of the obtained solution. The user can visualise the solution within the GIS planning system and make the changes if necessary.

9.7 Worked Examples

In this section, an example of using the design tool to plan two different networks is presented. *Net1* which contains 440 nodes and *Net2* which contains 1,163 nodes are shown in Figs. 9.4 and 9.5, respectively. Additional information of both networks is given in Table 9.1. For clarity, the background of plot layout is removed. The nodes are connected by links which represent possible routes for cabling. Not all of these links will be used in the final design as a tree network solution needs to be formed. The entry point to the network (i.e. POP) or the point where the new network is connected to the existing one is represented by a red box. The blue squares are the FDHs, the green ellipses are the possible locations of DPs, and the black squares represent the entry nodes which are used to connect to optical network units (ONU)

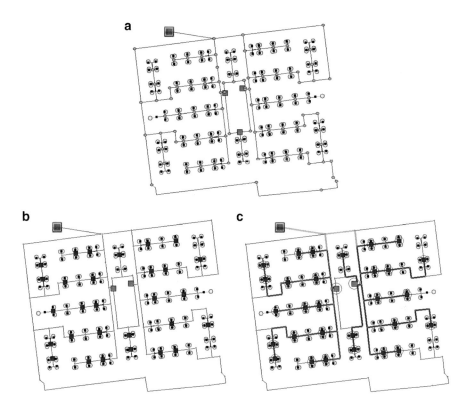

Fig. 9.4 Network *Net1* with design results

inside the customer premises (yellow nodes). Each ONU may require multiple connections.

The resulting network design will greatly depend upon the parameters shown in Table 9.2. They can be categorised in two types. The first one refers to capacity constraints imposed by the hardware limitations of the network equipment being used. These constraints may be changed due to the advancement of new hardware. The second type of parameters deals with distance constraints which ensue from the signal loss. The signal loss is caused by fibre connections and attenuation, splices, optical signal splitting, etc. In GPON, the power budget, measured in decibels (dB), is defined as the difference between the maximum transmitting power of the OLT and the minimum receiver sensitivity at the ONU located in a customer premise. As the number of connections and splices along the path from a customer to an OLT, in general, is determined by the topology being used, the power budget constraints are transformed into distance constraints. Distance constraints are often used to limit the coverage of network equipment for the purpose of network management.

Regarding the optimised result, the solution network for *Net1*, taking approximately 10 n, is shown in Fig. 9.4b, c. Figure 9.4b shows the DP locations represented by purple squares, while Fig. 9.4c shows the cabling solution rooted

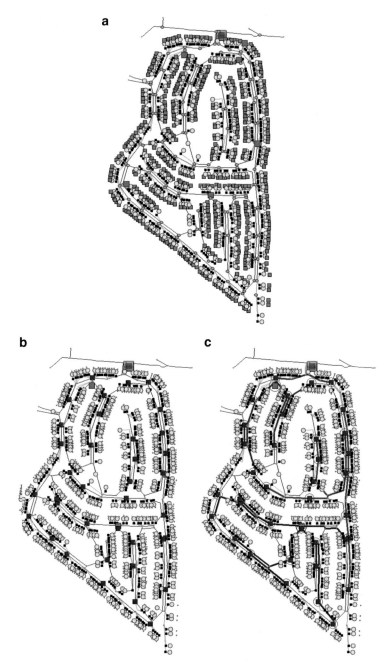

Fig. 9.5 Network *Net2* with design results

Table 9.1 Summary of input information and design results

	Customer nodes	Customer Demand	Possible DPs	Possible FDHs	Selected DPs	Selected FDHs
Net1	214	314	93	3	29	2
Net2	506	523	105	6	37	3

Table 9.2 Design parameters

Planning Parameters	
Max no. of connections per splitter	32
Max no. of connections per DP	16
Max no. of DP per FDH	16
Max distance between a DP and FDH	1600 meters
Max distance between a customer and DP	350 meters

from the entry point. A feeder cable which usually contains a few hundred fibres runs from the entry point of the network to each FDH. At the FDH, fibres are branched out to feed DPs to which the customers are connected using drop cables. A drop cable can be a single core or multicore fibre cable depending on the demand of each customer. It usually has more fibres than the number of required connections for future use.

In addition, other practical criteria need to be taken into account, for example, a clear demarcation among FDH catchment areas to facilitate network installation and maintenance. The required demand of a given customer should be served by the same DP.

A more complex network *Net2* was used to test the performance of the optimisation process. The corresponding solution is shown in Fig. 9.5. In this case, the network has 506 customer premises. It took approximately 100 s to generate the solution, running on a 32-bit Windows-based laptop with an Intel Centrino2 vPro 2.4GHz processor and 4GB of RAM. Figure 9.5b shows the locations of DPs. In Fig. 9.5c, the blue, dark blue and red paths indicate the catchment areas of different selected FDHs.

Compared with the manual process, each FDH which can have the maximum of 256 connections (i.e. 16 connections per DP × 16 DPs per FDH) would normally take approximately 2 to 3 days for a planner to come up with a design that satisfies all the planning criteria. The process requires the planner to identify the FDH locations for the given network, form clusters of DPs within each FDH and then assign customers to each selected DP. The time required for manual design can vary depending on the size and the layout of the given network. For example, *Net1* shown in Fig. 9.4 is much more symmetrical than *Net2* in Fig. 9.5. Therefore, it could take significantly more time for the planner to create the network design for *Net2* than *Net1*. Furthermore, with the optimisation tool, planners can specify more locations of DPs and FDHs than actually required. The tool will automatically choose the optimal or near-optimal ones. Apart from the time saving, the benefits of auto-generation of cabling diagrams and bills of material (BOM) can be also achieved through the automation of network design process.

In addition, an important aspect of the optimiser observed from the solution is the sensitivity to the ratio of the network equipment cost to the cable cost. Both costs include component (i.e. FDH, DP, cable) and labour cost (i.e. installation cost). When the ratio is high, the optimiser tends to use longer cables and populate less DPs to centralise the customers. In this situation, a good usage of the network equipment can be achieved. However, when the cable cost outweighs the equipment cost, more network equipment would be deployed. The overprovisioning of network equipment may be good for future growth. However, if too many spare connections are left unused for a long time, the return on investment will be delayed.

9.8   Conclusions

Many researchers are focusing on improved algorithms for a particular problem. Down-streaming a piece of research and development work to solve practical network problems is of equal importance. However, accomplishing this task was very challenging. It required a tremendous amount of effort and experience and involved many engineering staff to formulate the problem in such a way that could be implemented. During the development stages, many amendments were made according to network planners' suggestions. In the final stage, detailed verifications of this tool from different aspects were undertaken to ensure the produced solutions not only reached a high standard of quality, but also that the optimisation module met the engineering and reliability criteria of the software specification.

This article provides a detailed case study of applying AI methods to produce efficient practical FTTH network design on Greenfield development sites. The short computational time required to produce a cost-effective and consistent solution while comparing the manual process confirms the advantage of employing the ACO approach to tackle this type of problem. The business benefits of applying the automated design methods are listed below:

- Guaranteed compliance of the resultant design to the network designers' defined design criteria and constraints
- Minimisation of network capital expenditure, construction and installation, and materials costs
- Significantly reduced network design time compared to manual design methods
- Reduction in the skill required by the network designer as the expert design rules have been embedded in the software module
- Ability to rapidly re-cost networks to meet changes in requirements
- Production of network design in electronic format, ready for uploading to central database

Further development will focus on processing efficiency improvement for larger-sized networks, which includes the combination of meta-heuristic methods and exact methods based on the Mixed Integer Linear Programming approach. In addition, we would like to extend our optimisation module to tackle Brownfield areas by taking into the existing network infrastructure and planning constraints.

References

Cambazard H, Mehta D, O'Sullivan B, Quesada L, Ruffini M, Payne DB, Doyle L (2011) A combinatorial optimisation approach to designing dual-parented long-reach passive optical networks. CoRR, vol. abs/1109.1231

Chu A (2009) Ant colony optimization metaheuristics for solving combinatorial optimization problems. University of Economics, Prague, Faculty of Informatics and Statistics

Dorigo M, Stützle T (2004) Ant colony optimization. The MIT Press, Cambridge

Jaumard B, Chowdhury R (2012) Selection and placement of switching equipment in a broadband access network. In: Computing, networking and communications (ICNC), 2012 International Conference, pp 297–303

Kim Y, Lee Y, Han J (2011) A splitter location/allocation problem in designing fibre optic access networks. Eur J Operat Res 210(2):425–435

Lakic B, Hajduczenia M (2007) On optimized passive optical network (PON) deployment. In: Access networks workshops, 2007. AccessNets '07. Second International Conference, pp 1–8

Li J, Shen G (2009) Cost minimization planning for Greenfield passive optical networks. IEEE/OSA J Opt Commun Network 1(1):17–29

Lv M, Chen X (2009) Heuristic based multi-hierarchy passive optical network planning. In: Wireless communications, networking and mobile computing, 2009. WiCom '09, 5th International Conference, pp 1–4

Mazzali C, Whitman R, Deutsch B (2005) Optimization of FTTH passive optical networks continues. Lightwave Magazine, January

Penrose MD (1995) Single linkage clustering and continuum percolation. J Multivariat Anal 53 (1):94–109

Poon, KF. and Ouali, A. (2011): "A MILP based design tool for FTTH access networks with consideration of demand growth", in Internet Technology and Secured Transactions (ICITST), 2011 International Conference, pp. 544 –549.

Poon K, Mortimore D, Mellis J (2006) Designing optimal FTTH and PON networks using new automatic methods. IET Conference Publications, vol. 2006, no. CP521, pp 45–48

Steve C (2008) Designing low cost access networks with IPTV performance constraints. In: Next Generation Internet Networks, NGI 2008, pp 45–52

TechNavio, Global Passive Optical Network Equipment Market 2011–2015, SKU: IRTNTR1566, http://www.technavio.com

Chapter 10
The Role of Service Quality in Transforming Operations

Gilbert Owusu, Paul O'Brien, and Sid Shakya

Abstract The introduction of any tool requires changes in the users' environment to use the tool. McAfee (Harv Bus Rev November: 141–149, 2006) highlights this point by stating that the challenges in IT projects are not just technical but managerial. Here, managerial refers to embedding the system within the organisation. Kotter says '*in the final analysis, change sticks when "it becomes the way we do things here", when it seeps into the bloodstream of the corporate body*' (Kotter, Harv Bus Rev January: 96–103, 2007). Clearly, the success of any IT transformation programme is in part a function of the quality of service being provided by the system. BT not only is a consumer of service and field automation technologies but also provides production management solutions to other industries. We have observed (both qualitatively and quantitatively) from our experiences of developing production management systems that the quality of the services being provided by production management systems is dependent on the perceptions of the users of the system. This correlates with the measures put in place for engagement between the development team and the end users during the life cycle of the development.

10.1 Introduction

The effective measurement and management of customer satisfaction provides an invaluable approach to improving service quality. Beach and Burns (1995) view customer satisfaction as comparing customer evaluations of the services they experience with the planned experience. The aim is to identify what customers see as *matching, falling short* or *exceeding expectations*. Kotler (1997) defines customer satisfaction as a person's feelings of pleasure or disappointment resulting from comparing a product's perceived performance in relation to his or her

G. Owusu (✉) • P. O'Brien • S. Shakya
Research and Innovation, BT Technology, Services and Operations, Martlesham Heath, UK
e-mail: gilbert.owusu@bt.com; paul.d.obrien@bt.com

G. Owusu et al. (eds.), *Transforming Field and Service Operations*,
DOI 10.1007/978-3-642-44970-3_10, © Springer-Verlag Berlin Heidelberg 2013

expectations. Parasuraman et al. (1985) view quality as a comparison between expectations and performance. Smith and Houston (1982) assert that satisfaction with services is related to confirmation or disconfirmation of expectations. The motivations for pursuing service quality are varied and many, including service improvement, profitability, customer satisfaction, customer retention and minimising operational costs (Reichheld and Sasser 1990; Johnston and Clark 2008). Silvestro et al. (1990) note that service organisations use customer satisfaction measures and internal and external data to measure service quality.

In this chapter, we focus on capturing end users' perceptions and customer insights with the view of improving quality of services provided by production management systems. Understanding and implementing customer's needs will improve the success of any IT-enabled transformation programme. Johnston (2009) notes that customer insight is about developing a clear understanding of customer's needs, expectations and perceptions. We use gap analysis and a variant of SERVQUAL (Parasuraman et al. 1994) for measuring the gap between customers' priorities and their perceptions of services provided by production management systems and the match between the customer[1] and the delivery team's perspective. The focus here is the development and delivery of a production management system in BT. We start with a brief description of the service operation in Sect. 10.2. In Sect. 10.3, we present a cursory review of the literature on the use of gap analysis to measure service quality. The methodology applied in the case study is outlined in Sect. 10.4. We present the research findings in Sect. 10.5. We highlight the strategies for improvement in Sect. 10.6 and draw out the lessons learnt in Sect. 10.7.

10.2 The Service Operation: Development and Delivery of a Production Management System in BT

The service operation discussed in this chapter is the development and delivery of software for managing BT's field engineers. The software enables resource managers to have full visibility of customer demand and the capabilities and the capacity of the field engineers to deliver services that customers have requested. The customers are the resource managers who manage the field engineers. The development and delivery teams are located in the UK and offshore. Typically, at the beginning of a financial year, the customers will supply a list of requirements to the delivery team. This list is then translated into software specifications and follows standard software development process of design, develop and test, user acceptance testing and the deployment into an operational environment. The software is also supported by the development and delivery teams[2] once it has

---

[1] Customer and end user are used interchangeably.

[2] We use the term delivery teams hereafter.

Fig. 10.1 The development
process

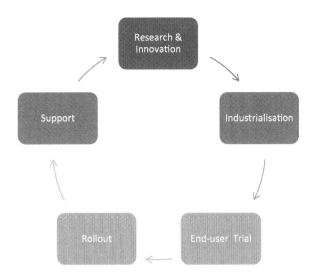

become operational. Figure 10.1 provides a schematic diagram of the development
process. One of the key challenges faced in any IT development and deployment
programme is ensuring that the software is fit for purpose and works right first time.
Defining quality measures early in the development cycle and seeking feedback
using prototype accelerates the uptake of any IT system. In the next section, we will
present some approaches for defining and capturing software quality measures.

10.3   Background

Despite the multiplicity of views on the similarities and differences between
definitions of service quality and customer satisfaction, Johnston (1995) observes
that there is a general consensus that the two are related. Silvestro et al. (1990) note
that service is usually the result of the interaction between the customer and the
service system. They also highlight that the provision of service quality is
concerned with generating customer satisfaction. Davis and Heineke (1994) argue
that customer satisfaction provides the linkage between the level of service that a
firm provides and the customer's perception of that service. Cronin and Taylor
(1992) observe that perceived service quality is a function of attitude. Johnston also
views service quality as the consistent conformance to specification. The most
widely held view is that service quality is the degree of fit between expectations
and perceptions (Parasuraman et al. 1985; Davis and Heineke 1994; Johnston
1995).

Since the early 1980s, a number of models have been developed to measure
service quality. These models focus on measuring the gap between customers'
expectations and perceptions of the service delivered (Beach and Burns 1995;

Silvestro 2005; Johnston 2009). We refer to these models as the gap analysis tools. Almost all the models are a derivative of SERVQUAL. SERVQUAL was developed by Parasuraman et al. (1994). It is primarily a multiple-item tool for assessing overall perceived quality. Other extensions to SERVQUAL have focused on measuring perceptions with the view of eliciting practical improvement priorities. For example, Silvestro's (2005) work in the health service employed a variant of SERVQUAL for measuring the gap between patient's priorities and their perceptions of an NHS service with the view of producing a service improvement agenda. Beach and Burns (1995) advocate the use of the 'Quality Improvement Strategy (QIS)' to measure service quality. Two types of gap analysis can be undertaken with QIS. The first examines gaps between expectations and perceptions of services offered to customers. The second focuses on gaps between services offered by an organisation and its competitors.

The question is *how does one quantify or measure quality*? Service delivery is multifaceted, and there is a widely held view that customer's expectations are rarely concerned with a single aspect of a service package. Rather customers tend to be interested in all the aspects of a service delivery. In essence these different aspects are attributes or factors of service quality. Identifying these factors is a prerequisite for addressing any gaps in service quality. Parasuraman et al. (1994) refer to these factors as the *dimensions* of service quality. Johnston (1995) refers to this as the *determinants* of service quality. Parasuraman et al. (1994) identify ten quality factors: access, communication, competence, courtesy, credibility, reliability, responsiveness, security, understanding and tangibles. Johnston (1995) extends Parasuraman et al.'s list to 18 quality factors: cleanliness, aesthetics, comfort, functionality, reliability, responsiveness, flexibility, communication, integrity, commitment, security, competence, courtesy, friendliness, attentiveness, care, access and availability.

Are these factors applicable to all service industries? Berry et al. (1985) contend that the ten quality factors by Parasuraman et al. are comprehensive and are applicable to all service industries. A number of studies have confirmed the applicability of the factors to service industries such as information systems (Jiang et al. 2000), health (Silvestro 2005), hotel (Fernándz and Bedia 2005) and telecommunications (Sattari 2007). A common thread running through these studies is to accelerate the uptake of IT systems and thus improve the success of related IT-enabled transformation programmes.

10.4   Methodology

The work of McCall et al. (1977) helped pioneer the use of quality factors for software. They identified 11 factors: efficiency, integrity, reliability, usability, accuracy, maintainability, testability, flexibility, interface facility, reusability and transferability. Boehm (1984) also produced 19 factors: usability, clarity, efficiency, reliability, modifiability, reusability, modularity, documentation, resilience, correctness, maintainability, portability, interoperability, understand ability, integrity, validity, flexibility, generality and economy. There have been variations of McCall's list, for example, lists produced by Murine and Carpenter (1984) and Ghezzi et al. (1991). It is worth noting that the lists produced by McCall et al. and Boehm predate Parasuraman et al.'s ten dimensions and Johnston's 18 quality factors. This suggests that researchers in IT/IS[3] always viewed the delivery of software as a service. There are commonalities to the lists produced by both service quality and IT researchers.

Berkley and Gupta (1995) suggest that asking selected customers to audit actual service delivery is a simple method for measuring customer reaction, thus providing the framework to elicit any factors that are important to customers. They list questionnaires, interviews and rating cards as examples of auditing systems. Based on interactions (i.e. review and retrospective sessions) with end users (i.e. customers) of the service operation outlined in Sect. 10.2, we have produced a list of 16 factors along the lines of the SERVQUAL methodology. The list is in line with the lists produced by Johnston (1995) and Boehm (1984). We list the statements which were used to capture each quality factor in Table 10.1. Two sets of questions were produced for the 16 factors. Thus, there were 32 questions in total. A survey was produced with these questions and sent to end users and the teams delivering the service. There are two types of end users: the 'actual' end users and end users' technical team. The actual end users use the service (i.e. software) on a daily basis. The end users' technical team on the other hand acts on the behalf of the end users ensuring that the business requirements have been implemented by the delivery teams. The first set statements related to the priorities of the end users on a 1–5 scale (one being unimportant, five being very important). The second set focuses on the perceptions of service delivery. End users were asked to assign a value along a five-point Likert scale (one being very poor, five being very good). The teams responsible for delivering the service were also asked to assign values to statements. The delivery teams come from three functional areas: *programme management*, *development team* and *end-to-end test teams*. The programme management team ensures that the software is developed and delivered to the customer within budget, on time and to specification. The development team is responsible for developing the software, whilst the test teams ensure that the quality of the software meets the customers' requirements. The survey was sent to 42 end users, 7 end users' technical personnel, 6 programme managers, 40 software developers and 5 testers. The responses were 13 for the end users, 4 for the end users' technical team, 4 for the programme managers, 11 for the software developers and 2 for the testers. For the priorities, the delivery teams were asked to rate what they 'think' is the end users' priorities. For the statements related to the perceptions, they were asked to rate what they 'think' the actual end user's perception was. The motivation here is to identify gaps between the end users and the delivery teams.

---

[3] Information Technology/Information Systems.

Table 10.1 Quality factors and their statements

Quality Factor	Statement Used
Functionality	Does the software application perform all the desired functions for which it was developed?
Reliability	Is the software application reliable in terms of results?
Usability	Is the software application user friendly
Efficiency	Is the software application efficient in terms of responses
Maintainability	Is it easy to find and fix a defect
Integrity	Does the software provide protection from unauthorized access
Portability	Can the software be transferred from one environment to another?
Flexibility	Is it easy to make changes required as dictated by the business?
Speed	Is the software delivered on time?
Cost	Is the software delivered to budget?
Technical expertise	Technical capability of the team
Learn ability	How quickly do team members pick new processes?
Understanding of requirements	Does the delivery team appreciate customer requirements?
Understanding of business process	Is there an appreciation for business processes?
Security	Is the software secure?
Proactive identification and management of risks	Is there proactive identification and management of risks

10.5   Research Findings

The findings presented in this section attempt to answer the question: 'what are the mechanisms for using IT transformation programmes to institutionalise change?' As we have noted previously, the quality of the production management system determines the extent to which the system is used and consequently addresses whether it has been embedded in day-to-day operations. The data (Tables 10.2 and 10.3) for the research findings are presented in the Appendix. Table 10.2 presents the research data. The values represent the mean across each factor for each 'stakeholder[4]'. In Table 10.2, we present priorities and perceptions of the end users, their technical team and the delivery teams. Table 10.3 highlights the priority-perception gaps for each factor and for the end users, their technical teams and the different delivery teams. We analyse the findings below:

- *End users' priorities*: All factors with the exception of 'portability' are rated highly for both end users and their technical teams. The technical teams gave a rating of 3 to 'portability'. This is expected since they are conversant with the technology[5] that is being used to develop the software. The mean importance level for the end users is higher (4.73) than their technical teams (4.31). This reflects the importance and value of service delivery to the actual end users. It is interesting to note that the technical teams gave the highest score—5, to functionality, efficiency, technical expertise, security, reliability and integrity. These are system features and what can be perceived as the hygiene factors for software delivery. The highest rating for the end users is for functionality, usability, technical expertise and understanding of business process. It is interesting to note that the end users focus more on the competence of the team and to some extent the softer side of the service delivery than their technical teams. This is understandable since a major concern for them is to get the appropriate support needed to use the service.
- *End users' perceptions*: With the exception of 'integrity' and 'understanding of business process' and 'security', all factors were rated below 3 by the end users. The lowest perception rating by both the end users and their technical teams was for reliability. Comparison of the perceptions of the end users and their technical teams raises some interesting issues. The end users' perception rating is significantly lower than their technical teams. The mean perception rating across all factors for the end user is 2.56, whilst that of the end users' technical team is 3.44. A review of support complaints with the delivery teams reveals that the end users were dissatisfied with the reliability of the service.
- *Measurement of the priorities/perception gaps*: In general the perception for both the end users and their technical teams is negative. There were negative perception gaps across all the factors for the end user. The worst being reliability. The worst perception gap for the end users' technical team is also

---

[4] End users, end users' technical team, programme managers, development team and test team.

[5] Java technology is used and it is portable to other platforms.

'reliability' and has the same score, i.e. -3, as the end users. However, 'portability' exceeded the priority rating for the technical team. It is interesting to note
that what is considered as 'systems-related' factors (e.g. functionality, reliability) manifested the widest priority-perception gaps, -2.9 and -3.0, respectively,
for the end users. However, the softer side such as 'learn ability' and 'technical
expertise' recorded the lowest priority-perception gaps. The end users were
more dissatisfied with the service than their technical teams. This issue should
be explored further since the development teams consider both groups as
'customers'.

- *Analysis of end users and delivery teams' perceptions*: Comparing the mean
  important rating across all factors reveals that the programme managers (4.2)
  best understand customer's priorities. This is followed by the development team
  (4.1), then the testing team (3.9). It is worth noting that the testing team is based
  offshore in India. The development team is also based offshore; however, there
  is a co-ordinator on-site who interacts with the customers. The programme
  managers are all based on-site (in the UK) and have most contact with the
  customer. They hold weekly status calls with the customer and act as the
  relationship managers for the delivery teams. Silvestro (2005) reported a similar
  observation in her work with an NHS trust, where staff with most patient contact
  demonstrated the best understanding of patient's priorities. In terms of perceptions, the development and test teams each returned 4 factors rated less than
  4, whereas the programme managers rated 10 factors below 4. Again, this
  confirms the observation that programme managers had a better understanding
  of the customer's priorities. The list for the lowest perception for the programme
  managers is similar to the end users' technical team. However, the end users'
  technical team had issues with the factors related to 'understanding of requirements' and 'understanding of business process'. This is an interesting point,
  since they work with the delivery teams to articulate the end users' requirements.
  They also sign off the deliverables on behalf of the customers.
- *Analysis of perceptions among delivery teams*: From the list of important factors
  in Table 10.2, it would appear that the development team lacked an awareness of
  what was important to the customer. It looks like they believed that the customer
  did not classify the factors into *hygiene, enhancing, critical* and *neutral*. They
  treated all factors as nearly equal. They had the least standard deviation[6] (0.17)
  compared to the programme managers (0.39) and the test team (0.62).

10.6   The Service Quality Improvement Strategy: Implications for the Development and Delivery Teams

In summary, the end users had negative perception of the service that was delivered.
In particular, they had issues with the reliability of the service. It would appear that
reliability is a critical factor (Johnston and Clark 2008). *What are the hygiene,*

---

[6] Across important factors.

enhancing, critical and neutral factors? An analysis of the data reveals that the end users treat all the factors as hygiene factors. They had a standard deviation of 0.16 across the important factors—the least among the groups surveyed. Clearly, what is required is a clear delineation of the important factors into the four different groups. This will ensure that the expectations are clearly defined for the delivery teams. Despite this apparent lack of clarity on the classification of the importance factors, the programme managers understood best the priorities of the customers.[7] It is also worth noting that the most important factor for the end users, i.e. 'technical expertise', was not the same for the technical teams. The technical teams rated 'functionality' as the most important factor. Clearly, an agenda for improvements was required to correct the negative perception of the service delivered. Beach and Burns (1995) recommend a prioritisation strategy since some gaps may require more resources than others. An initial step was to expose the offshore teams to the customers. This was done via briefing sessions where customers (i.e. end users) provided regular feedback. Such an approach enabled the offshore teams to appreciate the priorities of the customers. Second, the issues related to reliability of the system were investigated. Was it a hardware or software problem? What were the service recovery mechanisms? We involved the customers in identifying answers to the above questions. Were there blockers to delivering the service? We put in place a plan for knowledge sharing among the delivery teams. We also addressed the perception gaps between delivery teams from the customers' perspective by having regular team briefings which focused on addressing the issues.

One major issue was the rotation of the offshore teams in the UK. We addressed this by ensuring that the offshore team mirrored the BT team. This way the impact of losing domain knowledge will be minimised. We also put in place more formalised training to new joiners on the offshore team. A typical training period for new joiners should include induction on both the technical and domain aspects of production management. We also realised that more effort should also be spent by the BT team on validating the knowledge of the offshore team. Here the congenital knowledge gained during set-up phase of any development is refined via a process of eliciting feedback. Huber (1991) refers to this process as experimental learning—i.e. the process of acquiring knowledge through direct experience. Working with the offshore team, the BT team also introduced weekly knowledge management sessions and quizzes with the view to declaring a knowledge management champion for the week.

10.7   Conclusions

We are continuously improving the way we develop production management systems. The systems we have developed have underpinned some of the major transformation programmes in BT. The underlying innovation development model

---

[7] This is reflected by the mean important rating across all factors.

has been recognised by leading organisations such as Global Telecoms Business, Professional Planning Forum, National Business Awards, National Outsourcing Association, EURO-INFORMS, UK Operational Research Society, BCS and IET. The systems are also being marketed externally to other utilities. Some important observations from this study are as follows:

- Berry et al. (1985) argue that regardless of the service being studied, reliability is the most important factor for service customers. This study has confirmed the above observation and highlighted the negative perception of the overall service if reliability is missing.
- The use of SERVQUAL to validate the classification of factors into hygiene, enhancing, critical and neutral factors. An analysis of the standard deviation of the ratings for important factors provides an indication whether the priorities have been classified.
- Berkley and Gupta (1995) note that service errors are often caused by a misspecification of the service and that quality in services depends heavily on the ability of employees to share their knowledge. We refer to this as the communication gap. There must be effective communication among delivery teams since the ability to deliver quality service depends on effective collection, processing and distribution of customer's requirements and priorities. Effective communication provides a mechanism to fine-tune the 'nonsystems related', i.e. softer, factors.

The case reported in this chapter demonstrated the use of an adaptation of SERVQUAL for measuring the gaps in the delivery of an IT service. Indeed there has been research on the use of SERVQUAL in IT services (Kettinger and Lee 1999; Jiang et al. 2000). One of the main limitations of SERVQUAL is that there is an assumption that customers remain stable during the whole analysis process. Van Dyke et al. (1999) outline other problems with using SERVQUAL as (a) the use of difference or gap score, (b) poor predictive and convergent validity, (c) the ambiguous definition of the expectation construct and (4) the unstable dimensionality.

Finally, we will also stress the need to achieve the right balance between knowledge management and cost containment: Managing IT investments in the light of recent outsourcing initiatives is a challenging task for most organisations. Organisational changes are typically addressed by changes in systems and processes. In recent years, there has been a drive to reduce IT spend in most organisations including governments. It is worth noting that the motivation for this drive is the perception that the development of software artefacts is similar to the process followed in the manufacturing sector. Manufacturing components are very well defined and in most cases can be sourced from multiple vendors. In software sector, software components are very much artefacts which require the developer's knowledge to maintain the systems developed. This fact seems to be ignored by IT budget holders, and there is the view that suppliers can be easily replaced. There is a high switching cost which can result in low-quality deliverables. There is also a lot of effort that goes into training the development teams whenever there are new joiners. More research is required to better understand the types of knowledge required for collaborative software development projects.

Appendix

Table 10.2 Research data

Priorities

	Functionality	Reliability	Usability	Efficiency	Maintainability	Integrity	Portability	Flexibility	Speed	Cost	Technical expertise	Learnability	Understanding of requirements	Understanding of business processes	Security	Proactive identification and management of risks	mean	sd
End users	4.9	4.8	4.9	4.8	4.7	4.7	4.8	4.7	4.8	4.3	4.9	4.7	4.7	4.9	4.6	4.6	4.73	0.16
Technical teams	5	5	4	4	5	4	5	4	4	4	5	4	4	4	5	4	4.31	0.60
Prog management	4.5	4.5	4.25	4	4	3.75	3.25	4.25	4.25	4.25	4.5	4	4.75	4.75	4	4	4.17	0.39
Dev team	4.2	4.2	4.2	4.2	3.9	4	3.9	4.3	4.2	4.2	4.4	4.2	4.2	3.9	4.2	4.4	4.14	0.17
Test team	4.5	3	4	4	3	4	4.5	3.5	3.5	4	4	4.5	4.5	4.5	4	2.5	3.88	0.62

Perceptions

	Functionality	Reliability	Usability	Efficiency	Maintainability	Integrity	Portability	Flexibility	Speed	Cost	Technical expertise	Learnability	Understanding of requirements	Understanding of business processes	Security	Proactive identification and management of risks	mean	sd
End users	2	1.8	2.2	2.2	2.2	2.1	2.9	2.2	2.2	2.2	2.9	2.8	2.8	3	3.3	2.7	2.56	0.53
Technical teams	3	2	4	4	4	3	4	3	3	3	4	4	3	3	4	3	3.44	0.73
Prog management	4	3.75	3.5	3.75	3.75	3.75	3.5	3.5	3.75	3.75	4.25	4	4.25	4	4	3.25	3.80	0.28
Dev team	4.3	4	4.2	4.2	4	3.9	3.9	3.9	4.2	4.3	4.3	4.3	4.3	3.8	4.2	4.3	4.12	0.18
Test team	3.5	3.5	4	4	4	4	5	4.5	4.5	4	4.5	4.5	4.5	4.5	4.5	3.5	4.16	0.47

Gap analysis

	Functionality	Reliability	Usability	Efficiency	Maintainability	Integrity	Portability	Flexibility	Speed	Cost	Technical expertise	Learnability	Understanding of requirements	Understanding of business processes	Security	Proactive identification and management of risks
End users	-2.9	-3	-2.7	-2.6	-2.6	-2.6	-1.7	-2.5	-2.6	-2.1	-2	-1.9	-1.9	-1.9	-1.3	-1.9
Technical teams	-2	-3	0	-1	-1	-1	1	-1	-1	-1	0	0	-1	-1	-1	-1
Prog management	-0.5	-0.75	-0.75	-0.25	-0.25	0	0.25	-0.75	-0.5	-0.5	-0.25	0	-0.5	-0.5	0	-0.75
Dev team	0.1	-0.2	0	0	0.1	-0.1	0	-0.4	0	0.1	-0.1	0.1	0.1	-0.1	0	-0.1
Test team	-1	0.5	0	1	1	0	1	1	1	0	0.5	0	0	0	0.5	1

Table 10.3 Priority-perception gap for each factor for all stakeholders

	End Users	End Users Technical Team	Programme Managers	Development Team	Testing Team
Importance/perceptions gaps	Functionality (-2.9) Reliability (-3.0) Usability (-2.7) Efficiency (-2.6) Maintainability (-2.6) Integrity (-1.1) Portability (-1.7) Flexibility (-2.5) Speed (-2.6) Cost (-2.1) Technical expertise (-2.0) Learn ability (-1.9) Understanding of requirements (-1.9) Understanding of business process (-1.9) Security (-1.3) Proactive identification and management of risks (-1.9)	Functionality (-2.0) Reliability (-3.0) Efficiency (-1.0) Maintainability (-1.0) Integrity (-1.0) Portability (1.0) Flexibility (-1.0) Speed (-1.0) Cost (-1.0) Understanding of requirements (-1.0) Understanding of business process (-1.0) Security (-1.0) Proactive identification and management of risk (-1.0)	Functionality (-0.5) Reliability (-0.75) Usability (-0.75) Efficiency (-0.25) Portability (0.25) Flexibility (-0.75) Speed (-0.5) Cost (-0.5) Technical expertise (-0.25) Understanding of requirements (-0.5) Understanding of business process (-0.75) Proactive identification and management of risk (-0.75)	Functionality (0.1) Reliability (-0.2) Efficiency (0.1) Maintainability (-0.1) Integrity (0.1) Flexibility (-0.4) Cost (0.1) Technical expertise (-0.1) Learn ability (0.1) Understanding of requirements (0.1) Understanding of business process (-0.1) Proactive identification and management of risk (0.1)	Functionality (-1.0) Reliability (0.5) Efficiency (1.0) Integrity (-1.0) Portability (1.0) Flexibility (1.0) Speed (1.0) Technical expertise (0.5) Security (0.5) Proactive identification and management of risk (1.0)

References

Beach LR, Burns LR (1995) The service quality improvement strategy. Int J Serv Ind Manag 6 (5):5–15

Berkley JB, Gupta A (1995) Identifying the information requirements to deliver service. Int J Serv Ind Manag 6(5):16–35

Berry LL, Zeithaml VA, Parasuraman A (1985) Quality counts in service too. Bus Horiz 28:44–52

Boehm B (1984) Software engineering economics. IEEE Trans Software Eng 10(1):4–21

Cronin JJ, Taylor SA (1992) Measuring service quality: a reexamination and extension. J Market 56:55–68

Davis MM, Heineke J (1994) Understanding the roles of the customer and the operation for better queue management. Int J Oper Prod Manag 14(5):21–34

Fernández MCL, Bedia AMS (2005) Applying SERVQUAL to diagnose hotel sector in a tourist destination. J Qual Assur Hospit Tourism 6(1 & 2):9–24

Ghezzi C, Jazayeri M, Mandrioli D (1991) Fundamentals of software engineering. Prentice-Hall, New Jersey, NJ

Huber GP (1991) Organizational learning: the contributing processes and the literatures. Organ Sci 2(1):88–115

Jiang JJ, Klein G, Crampton SM (2000) A note on SERVQUAL reliability and validity in information system service quality measurement. Decis Sci 31(3):725–744

Johnston R (1995) The determinants of service quality: satisfiers and dissatisfies. Int J Serv Ind Manag 6(5):53–71

Johnston R (2009) Lecture notes. Service operations management. Warwick Business School

Johnston R, Clark G (2008) Service operations management: improving service delivery. FT/Prentice Hall, Harlow

Kettinger WJ, Lee CC (1999) Replication of measures in information systems research: the case of IS SERVQUAL. Decis Sci 30(3):893–899

Kotler P (1997) Managing service businesses and product support service. Prentice Hall, New Delhi

Kotter JP (2007) Leading change: why transformation efforts fail. Harv Bus Rev January; 96–103

McAfee A (2006) Mastering the three worlds of information technology. Harv Bus Rev November; 141–149

McCall JA, Richards PK, Walters GF (1977) Factors in software quality, vol I–III. Air Development Centre, Rome, Italy

Murine G, Carpenter C (1984) Measuring software product quality. Qual Progr 7(5):16–20

Parasuraman A, Zeithaml VA, Berry LL (1985) A conceptual model of service quality and its implications for future research. J Market 49:41–50

Parasuraman A, Zeithaml VA, Berry LL (1994) SERVQUAL: a multiple-item scale for measuring consumer perceptions of service quality. J Retail 64(1):5–6

Reichheld F, Sasser WE (1990) Zero defects: quality comes to services. Harv Bus Rev 68 (5):105–111

Sattari S (2007) Application of disconfirmation theory on customer satisfaction determination model in mobile telecommunication: case of prepaid mobiles in Iran. Master's Thesis, Lulea University of Technology

Silvestro R (2005) Applying gap analysis in the health service to inform the service improvement agenda. Int J Qual Reliab Manag 22(3):215–233

Silvestro R, Johnston R, Fitzgerald L, Voss C (1990) Quality measurement in service industries. Int J Serv Ind Manag 1:54–66

Smith RA, Houston MJ (1982) Script-based evaluations of satisfaction with services. In: Parasuraman A, Zeithaml VA, Berry LL (1994) Reassessment of expectations as a comparison standard in measuring service quality: implications for further research. J Market 58; 111–124

Van Dyke TP, Prybutok VR, Kapeelman LA (1999) Cautions on the use of SERVQUAL measure to assess the quality of information systems services. Decis Sci 30(3):877–891

Chapter 11
Field Force Management at *eircom*

Feargal Timon and Attracta Brennan

Abstract The case study outlined in this chapter focuses on *eircom*, Ireland's largest provider of telecom services, i.e. fixed line phone, mobile, broadband and data. The majority of *eircom's* technology business involves the delivery and repair of copper/fibre telephone services. Within that, *eircom's* field force comprises the technicians who install and repair these services. The management of *eircom's* field force has many challenges ranging from geography and skill to productivity and service level.

Simulation modelling was adopted as the most accurate approach to address these challenges. This chapter demonstrates how simulation was used to objectively understand the issues, design solutions and quantify *eircom*'s requirements. CIM Ireland built the models and assisted in the design of the solutions that resulted in a 16 % improvement in 'delivery on time', combined with a 19 % reduction in resources required. The accuracy level of the model's performance was 97.5 % when compared with actual performance.

11.1 Introduction

A field force refers to workers based in different geographic locations who travel (outside the office) *in the field* to their allocated tasks. Each worker has a set of skills and each task has a skill requirement. The tasks can be categorised as being either planned or reactive, e.g. the installation of phone services or the repair of phone lines.

Accordingly, field force management refers to the process of allocating tasks to appropriate workers who have the required skills in the area. It also involves the

F. Timon (✉)
CIM Ireland, Galway, Ireland
e-mail: feargal.j.timon@simulation.ie

A. Brennan
GMIT, Mayo, Ireland

G. Owusu et al. (eds.), *Transforming Field and Service Operations*,
DOI 10.1007/978-3-642-44970-3_11, © Springer-Verlag Berlin Heidelberg 2013

movement of technicians into areas with backlog/capacity requirements. The aim is to optimise the scheduling and dispatching of tasks to balance competing demands, worker productivity and completing tasks *on time* (Lesaint et al. 2000). Service level defines how *on time* is measured. Field force management is vital as 'one of the cornerstones of successful organisations has been the optimal use of their workforce' (Owusu et al. 2008). In this chapter we outline a case study using simulation techniques to transform eircom's field force. The chapter is structured as follows: We begin by outlining the factors that affect field force management. We highlight eircom's field force challenges in Sect. 11.2. Section 11.3 describes the simulation approaches we employed to address the challenges. In Sect. 11.4 we describe the operational impact of the transformation programme. We then briefly provide a perspective on other approaches and solutions for field force management in Sect. 11.5.

11.1.1   The Factors Affecting Field Force Management

The core challenge faced by field force management is to balance service level with productivity while taking into account skills and geography. The factors that influence this challenge include (Mohamed et al. 2012; Kern et al. 2009):

- **Density variation**: Tasks and workers are dispersed geographically. The typical time to travel to the task can vary based on customer density (e.g. rural/urban).
- **Task variation**: The task type, task duration and skill requirements are dependent on the product/location.
- **Demand variation**: The volume of tasks/demand in each area can vary significantly on a daily basis.
- **Hard boundaries**: This is where technicians will travel large distances within their own team boundary but are restricted from travelling outside this boundary. Hard boundaries are a legacy of historical management structures and limit resource flexibility.
- **Planned[1] vs. reactive[2] work**: The challenge is to be able to adapt resources to meet reactive work without significantly impacting the delivery of planned work.
- **Local knowledge**: Knowledge of the local customers, geography and location of the equipment is required.

---

[1] Planned work refers to jobs where a customer **does not have** an immediate requirement. These jobs can therefore be scheduled based on resource availability (e.g. installation work and/or maintenance work).

[2] Reactive work refers to jobs where a customer has an immediate requirement. These jobs need to be completed within a contracted time frame (e.g. faults).

These factors combine together to impact the overall challenge for field force management.

11.1.2   Core Challenge: Balancing Service Level and Productivity

The core challenge of field force management is how to manage the above factors to balance service level and productivity. This can be achieved by defining business rules (in a dispatch tool) for the prioritisation of jobs, the areas in which technicians can work and travel limitations. To illustrate this more clearly, Table 11.1 demonstrates the following two scenarios:

- Where technicians only work in their home/primary area focussing on productivity
- Where technicians complete jobs based on due date, outside their home/primary area but within travel limitations (with a focus on service level)

In Table 11.1, Scenario 1 ignores due date and limits travel to the technician's home area. This gives a service level of 75 % and a productivity level of 100 %. In the second scenario, due date is prioritised and travel limitations are relaxed. This results in a service level of 81.3 % and a productivity level of 88 %.

Although this is a basic example, similar challenges are replicated **for every order** throughout the day in real-life field force management.

11.2   *eircom*'s Field Force Challenges

At the beginning of this process, *eircom*'s field force challenges were:

- **Density variation**: In 2008 *eircom* had approximately 1,200 technicians spread over 70,000 km^2 with an average technician covering 77.8 km^2 from 5.6 km^2 in Dublin to 116 km^2 in Donegal.
- **Task variation**: There were about 60–70 different tasks and skills required, ranging from setting up a broadband service (0.5 h) to bringing a cable to a rural property (12 h).
- **Demand variation**: During storms, reactive work (i.e. faults) more than doubled the normal volume (see Fig. 11.1), while requests for planned installations varied by 20 % from 1 week to the next.
- **Hard boundaries**: Each technician only worked within their team area.
- **Planned vs. reactive work**: *eircom*'s management approach was to prioritise reactive work over planned work. This resulted in unpredictable lead times for their installation work.

Table 11.1 Productivity and service level scenarios

	Scenario 1 Focus on Productivity			Scenario 2 Focus on Service Level		
Base Data	Location A	Location B	Total	Location A	Location B	Total
Jobs due Day 1	8		8	8		8
Jobs due Day 2		8	8		8	8
Number of technicians	1 (Tech A)	1 (Tech B)	2	1 (Tech A)	1 (Tech B)	2
Job duration (exc. travel)	2 Hours	2 Hours	NA	2 Hours	2 Hours	NA
Travel time A to B / B to A	1 Hour	1 Hour	NA	1 Hour	1 Hour	NA
Tech Info	Tech A	Tech B	Total	Tech A	Tech B	Total
Hours per day	8 Hours	8 Hours	16 Hours	8 Hours	8 Hours	16 Hours
Day 1	Tech A Tech B	Tech A Tech B	Total	Tech A Tech B	Tech A Tech B	Total
Completes	4 Jobs	4 Jobs	8 Jobs	4 Jobs 3 Jobs		7 Jobs
Unproductive/travel time				2 Hours		
Job Performance	Location A	Location B	Total	Location A	Location B	Total
Servicel Level	100%	100%	100%	100%	0%	100%
Tech Performance	Tech A	Tech B	Total	Tech A	Tech B	Total
Productivity	100%	100%	100%	100%	75%	88%
Day 2	Tech A Tech B	Tech A Tech B	Total	Tech A Tech B	Tech A Tech B	Total
Completes	4 Jobs	4 Jobs	8 Jobs	1 Job	2 Jobs 4 Jobs	7 Jobs
Unproductive/travel time				2 Hours		
Job Performance	Location A	Location B	Total	Location A	Location B	Total
Servicel Level	0%	100%	50%	0%	100%	86%
Tech Performance	Location A	Location B	Total	Tech A	Tech B	Total
Productivity	100%	100%	100%	75%	100%	88%
Overall						
Job Performance	Location A	Location B	Total	Location A	Location B	Total
Servicel Level of **all** jobs	50%	100%	**75%**	88%	75%	**81.3%**
Jobs remaining				0	2	2
Tech Performance	Tech A	Tech B	Total	Tech A	Tech B	Total
Productivity	100%	100%	**100%**	88%	88%	**88%**

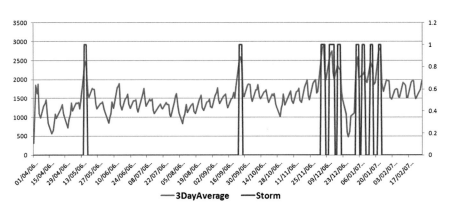

Fig. 11.1 Three-day rolling average of faults arrivals over a year

- **Local knowledge/work locally/technician loading policy**: Each day technicians were overloaded with the work in their area.

Technicians would complete jobs based on their own professional opinion as to what a customer needed and not based on eircom's business priorities. eircom's management of these challenges negatively affected the performance of the field force.

11.2.1 Impact on eircom's Field Force Performance

eircom's performance was affected by hard boundaries and overloading technicians and resulted in:

- **Fault service level** between 2006 and 2008 was at 60–65 % of faults completed in 2 days. The regulator's target was/is 80 %. The CEO had to appear on television programme (Prime Time, 2007) in response to the high level of complaints. See Fig. 11.2.
- **Incorrect sizing**: *eircom* was significantly overstaffed, but the technicians were in the wrong locations.
- **Overtime spend** was approximately €7 m per annum as a result of the incorrect sizing (see Fig. 11.3). The target for overtime was €3.5 million (Fig. 11.3).

The overall problem of being incorrectly sized (too many people, too little people in other places and hard boundaries) needed to be addressed (see Fig. 11.4). Also, the business rules used to manage the field force needed to be redesigned and evaluated. *eircom* needed to objectively address all of these issues, and therefore a new approach was required.

11.3 Why Simulation Modelling?

eircom's ability to predict staff requirements that would meet service level or design/evaluate business rules was too complex for spreadsheets due to the complications of (1) geography, (2) varying demand and (3) overtime. Meanwhile, Operations Research (OR)—a formulaic methodology (HSOR.ORG 2011)—had been tried as an approach but had not been found to be accurate. Table 11.2 evaluates the criteria for selecting an approach to resolve these issues. It was the business approach to justify the building of a dynamic simulation model.

11.3.1 The Simulation Model

The type of modelling used was discrete event simulation (Chahal and Eldabi 2008). Discrete event simulation 'models the operation of a system as a discrete sequence of events in time. Each event occurs at a particular instant in time and marks a change of state in the system'.

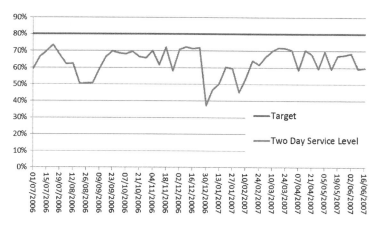

Fig. 11.2 Percentage of faults cleared in two working days

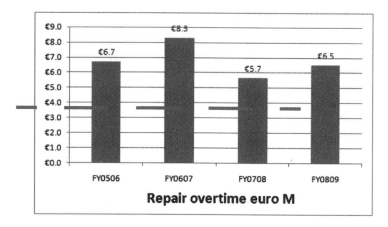

Fig. 11.3 Cost of overtime by financial year (FY)

Simulation in field force management refers to the simulation of the job's 'life' from creation and dispatch into the field (based on geography, task duration and travel time) to completion. Simulation (provided by CIM Ireland) was selected as it allowed *eircom* to:

- Evaluate and quantify the impact of changes before implementation.
- Quantify the number of technicians required in each location to maintain service levels and/or achieve required service levels.
- Cater to differing job requirements, e.g. the Dublin area was characterised by a more steady job flow and a smaller geographic area compared to the West coast, which had a highly variable number of faults and a larger geographic area.

CIM Ireland built and ran the models and was integral to the design of proposed solutions.

Fig. 11.4 Hard boundaries—technicians worked within the boundaries marked by the thick *red* and *blue lines*

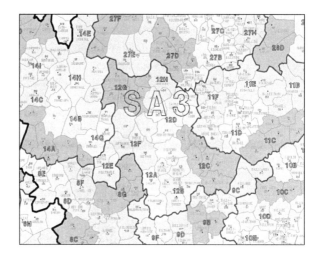

Table 11.2 Criteria used for the selection of a dynamic simulation approach to address field force planning issues

Factors	Spreadsheets	Operations Research	Dynamic Simulation
Hard Boundaries	Yes	Based on Standard Formulas	Yes/Based on Actual
Soft Boundaries	With difficulty	No	Yes/Based on Actual/Planned
Geography	Yes but not with other factors	NA	Yes/Based on Actual/Forecast
Varying Demand	To a degree	Based on Standard Formulas	Yes/Based on Actual/Forecast
Multiple Products	Yes but not with other factors	Yes but not with other factors	Yes/Based on Actual/Forecast
Skills	Yes but not with other factors	Yes but not with other factors	Yes/Based on Actual/Forecast
Impact of Overtime	Based on assumptions	Based on assumptions	Yes/Based on Implementable Rules
Output	Numeric	Numeric	Demonstratable/Validatable
Predict FTE requirement to meet Service Level	No	Not Accurately	Accurately
Design/Evaluate Business Rules	No	Not Accurately	Accurately
What if Analysis	Not Accurately	Not Accurately	Accurately

11.3.2 Model Overview

The model can be described as follows:

- Jobs (in the form of orders and faults) arrive into the simulation model, at an **exchange**[3] **level** and at the time and date at which they were originally created. This is referred to as the *arrival pattern*. Data used in this model is based on a

---

[3] Exchange—a telephone exchange links the cables from houses/businesses to the phone network. There are over 1,200 *eircom* exchanges throughout Ireland representing the geographic spread of the workload. This is the lowest level of granularity available to identify the location of jobs in *eircom's* network.

historical arrival pattern. The model run starts with a backlog of jobs as in reality.

- The model takes the arrival pattern and *adjusts* it to **match** *forecast volumes* (e.g. the forecast volumes for the next three business years were used).
- The time to complete a job is based on *past performance* plus a percentage productivity improvement year on year.
- Technicians are available for normal working hours (e.g. Monday to Friday 9 a. m. to 5 p.m.). Holidays, training days and non-productive time are dealt with by making the technicians unavailable. This varies throughout the year to represent different holiday/training requirements.
- Jobs sit in the model queue until they can be dispatched to an available technician (as in reality). The dispatch logic in the model imitates the logic of current and/or proposed dispatch tools.
- The dispatching logic (mainly used) takes jobs in priority and looks for a technician with the required skills and with the available capacity, who is based in the same area. If none is found, the logic then looks in neighbouring areas. Versions of the dispatch logic have been developed which use:

  - x–y co-ordinates
  - Travel by road
  - The use of areas and neighbouring areas in general to select the technician

  Note: next-generation dispatch logic/tools are outlined in Sect. 11.5.

- The technicians complete the jobs in their queue, based on priority rules (e.g. earliest due date, priority or closest job).
- When the jobs are completed by the technician, the service level for that job is recorded.

This model allows changes to the following inputs to be simulated:

- Job volumes at exchange level
- Job durations by area
- Addition/removal/movement of technicians at area level
- Dispatch logic/prioritisation

The impact of such changes on (1) service level, (2) technician utilisation and (3) overtime required to be quantified at any level of granularity is also simulated. The model accuracy can predict service level to within 2.5 % over a year and up to within 1 % within a month. Figure 11.5 represents the flow of information in the model, while Fig. 11.6 shows the model overview.

A sample scenario is where there is an anticipated productivity improvement. This could increase the potential service level beyond the requirements of the business as they would be overstaffed. The question for the simulation model is:

> What impact will this have on the number of technicians required, while achieving the target service level?

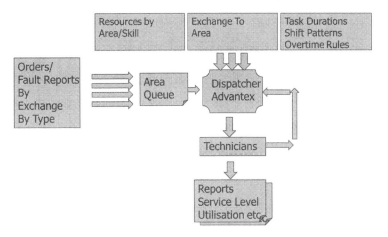

Fig. 11.5 Overview of model logic

Inputs	Simulation Model	Outputs
• Volume forecast • Arrival pattern • Unit cost/mhrs • Business rules • Skills	• Simulation (Witness) • 'What if' scenarios	• Resources required • Utilisation • Service Level • 'What if' scenarios

Fig. 11.6 Overview of the simulation model

In this situation, technicians are removed (in the model at an area level), the model is run and a service level is recorded. This process is repeated until the service level is brought back to match business requirements. Other areas where the model can be used include:

- The impact of soft boundaries
- Defining the areas in which technicians should work
- Business rules to balance productivity and service level, e.g.:

  – Busy areas may focus more on productivity.
  – Less busy areas may focus on service level.

- Impact of investment in the network
- Changes in how jobs should be dispatched
- Defining when to turn on overtime

Simulation can objectively evaluate solutions to these issues and accurately size the solution. Simulation can also be used in the effective management of competing demands (i.e. reactive vs. planned work)—an often difficult aspect of field force management.

11.3.3   How to Manage Competing Demands

There are three approaches to managing competing demands:

1. **Complete all planned work regardless of reactive work:** This does not take advantage of the flexibility of resources.
2. **Complete reactive work at the expense of planned work**: The service level of planned work can be seriously affected.
3. **In normal volumes, planned work is completed ahead of schedule leaving flexibility to move resources to reactive work when volumes increase**: It is important to quantify the duration at which resources can be moved to reactive work to maintain service level. As an example, assume you have products that have a 10-day lead time. If in normal volumes, these are completed within 5 days; this leaves 5 days where resources on planned work can move over to reactive work without affecting the service level of planned work.

In most companies (of which the author is aware), option three is not designed, quantified or sized correctly. Workload triggers need to be quantified at an area level to identify how many staff are required and for how long they can be redirected. Simulation modelling is the most accurate and effective approach to addressing all of the above. The simulation model was used to help *eircom* design their response to their field force challenges.

11.4   eircom's Response

eircom's response was based on simulation model outputs but also took process/cultural factors into account. The main changes were:

- As an interim solution to managing competing demands, work was split into the following two business units in order to improve business processes:

  - Delivery (planned work)
  - Repair (reactive work)

- The control of work management was centralised.
- The country was split into 244 areas and 40 teams.
- The number of technicians required by area was calculated by the model and implemented as part of the change programme (model output).
- In the *eircom*'s dispatch tool, all areas were linked to their neighbouring areas (i.e. secondary areas) (model output).
- Jobs were ordered based on priority rules that mainly focussed on service levels (i.e. earliest requirement date).
- The dispatch tool would dispatch a job based on a technician's home/primary area first and then look for jobs in their neighbouring/secondary areas even if this was outside the team area—this eliminated hard boundaries (model output).

- The use of primary and secondary areas gave rise to a *domino effect*. As an example, if there was a high level of faults in the north of the country, technicians would all move one step in the direction of the work.
- National response teams were established. These represent mobile technicians who move on a daily/weekly basis to areas where required (model output).
- Overtime rules (i.e. *trigger points*) were calculated by the model in order to ensure that overtime was turned on early enough to help improve service level.
- Weekly capacity planning meetings were set up in order to move technicians to areas with backlog.
- Local knowledge issues were addressed by:

  - Using IT systems to localise equipment and GPS to reduce travel times
  - Moving technicians into different areas to remove the reliance on an individual technician

- Responsibilities were split between *field* and dispatch *centre* (see Table 11.3).

After correctly sizing the organisation and implementing soft boundaries using the dispatch tool, *eircom* saw a significant improvement in the following:

- **Field resources**: reduced by 19 %.
- **Service level**: increased by 16 %. See Fig. 11.7.
- **Productivity**: remained constant.
- **Overtime**: decreased by 60–80 %, depending on year. See Fig. 11.8.
- **Savings**: of approximately €6.8 m.
- **NPS** (Net Promoter Score): up 14 points between March 2011 and January 2012.

The focus of *eircom* (from senior management to technicians) is now on service level

11.5 How Is *eircom* Unique in Its Use of Simulation?

The success of using simulation modelling and its ability to be objective (i.e. not biased to manager, union, board[4] or customer) has resulted in it being a trusted tool within *eircom*. As an example, resource deployment changes are modelled with the results being presented to the union before decisions are implemented. Similarly, if the board or the regulator[5] proposes a business change (e.g. a change in working hours), the model is used to evaluate the impact. Simulation, which is now an integral part of *eircom's* business/strategic planning, is used as an arbiter (see Fig. 11.9).

---

[4] The board of management makes strategic decisions for the company.

[5] The government appointed a regulator who defines the level of services to be achieved and can levy fines if these service levels are not achieved.

Table 11.3 Field and centre responsibilities

Centre responsibilities	Field responsibilities
Scheduling, prioritization and word distribution	Resource availability
Jeopardy management, escalations	Production/productivity
Resource planning	People management
Resource utilization	Quality
Service level agreement management	Health & safety

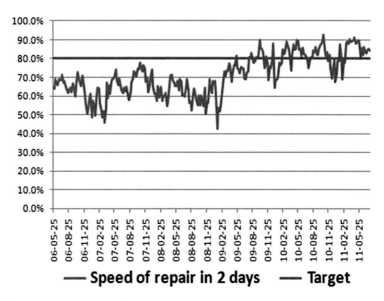

Fig. 11.7 Two-day speed of repair

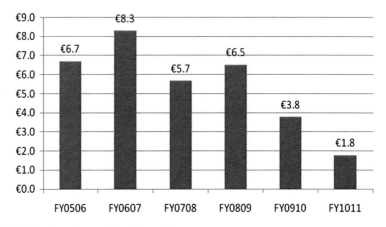

Fig. 11.8 Overtime by year (in millions €)

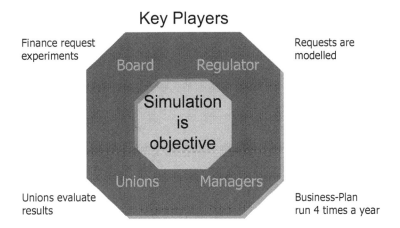

Fig. 11.9 Key players—simulation as an integral/objective part of *eircom*

11.5.1 Dispatching the Next Generation

eircom's implementation of the simulation model's recommendations (case study and current use) has identified weaknesses/requirements in their dispatch tool. Other field force managers have had similar experiences with their dispatch tools. From an initial review of next-generation dispatch tools on the market, the following are the key new features:

- **Dynamic road-based routing**: calculates the travel time by road from point A to point B, taking traffic into consideration (e.g. Bing, Google Maps, PTV (PTV 2013), PG Routing (PGRouting 2013)). This can use 2 approaches:
  - Routing is optimised *after* the technician is allocated the jobs.
  - Routing is optimised *before* the technician is allocated the jobs.
  - Approach 2 is the preferred option as it groups jobs, based on shortest travel time, into a bundle before giving them to the technician.

- **Optimisation**: refers to the running of multiple schedules and the scoring of each of these schedules based on their performance, in terms of productivity and service level. An effective optimisation tool uses algorithms to minimise the number of schedule runs while achieving an optimum outcome. Examples of optimisation scheduling tools include TOA, CLICK, Scheduling 360, Service Power and BT.

- **Artificial intelligence (AI)**: AI support tools offer solutions that assist managers or technicians in making decisions (e.g. recommendation of areas which require over time) or actually make the decision itself (e.g. automatically contact the appropriate technician to request that they work over time).

- **A more effective ability to balance productivity with service level**: the ability to easily adjust the business rules for different operating environments, for example, during high volumes (i.e. storm mode), the business rules may focus on productivity, while during normal volumes (i.e. normal mode), the business rules may focus on service level.
- **Learning-based algorithms**: refer to systems that learn how long a technician takes to carry out a task and, based on this, loads them with tasks appropriately.
- **Cloud-based systems**: 'Cloud computing describes a variety of different computing concepts that involve a large number of computers that are connected through a real-time communication network (typically the internet)' (Mariana Carroll 2012). In cloud-based dispatch tools, the full application and all data are stored on the cloud in both public and private domains.

While the Gartner report on field force dispatch evaluates current dispatch tools (McNeill et al. 2012), the authors are currently involved in identifying how effectively dispatch tools operate in actual organisations (i.e. do they work?). This ongoing work compares and contrasts dispatch tools based on (1) sample orders, (2) how they are being used by organisations and (3) whether or not they are meeting organisational needs.

11.6   Conclusions

This chapter discusses part of *eircom*'s ongoing journey to improve customer experience and reduce cost. *eircom* was in difficulty with service levels ranging between 60 and 64 % in 2007 and a belief that they were nationally overstaffed. Due to the complex interactions of skill, geography and competing demand of planned and reactive work, *eircom* chooses simulation modelling (provided by CIM Ireland) as the most accurate approach to design their solution. A simulation model was used to design a new field force organisation, size technician requirements at an exchange level and design the business rules to manage the field force.

This resulted in a 16 % improvement service level, a 19 % reduction in technicians and overtime going from 7 million to 3.5 million per year. Simulation (provided by CIM Ireland) continues to be used within *eircom* as an arbiter to evaluate business decisions. And as part of this journey to improve customer experience and reduce costs, new requirements for a dispatch tool have emerged (to further balance service level and productivity). CIM Ireland is currently exploring new tools on the market.

References

Chahal K, Eldabi T (2008) Applicability of hybrid simulation to different modes of governance in UK healthcare. In: Mason SJ et al (eds) Proceedings of the 2008 winter simulation conference. pp 1469–1477

Kern M, Shakya S, Owusu G (2009) Integrated resource planning for diverse workforces, computers and industrial engineering. In: Proceedings of the IEEE international conference on computers and industrial engineering, Troyes. pp 1169–1173

Lesaint D, Voudouris C, Azarmi N (2000) Dynamic workforce scheduling for British telecommunications plc. Interfaces 30(1):45–56

McNeill W, Maoz M, Van Huizen G (2012) Magic quadrant for field service management. http://www.gartner.com

Mohamed A, Hagras H, Shakya S, Owusu G (2012) Tactical resource planner for workforce allocation in telecommunications. In: Proceedings of the 2012 international conference on autonomous and intelligent systems, Aveiro, Portugal. pp 87–94

Owusu G, Anim-Ansah G, Kern M (2008) Strategic resource planning. In: Voudouris C, Owusu G, Dorne R, Lesaint D (eds) Service chain management: technology innovation for service business. Springer, Germany

Chapter 12
Understanding Team Dynamics with Agent-Based Simulation

Thierry Mamer, John McCall, Siddhartha Shakya, Gilbert Owusu, and Olivier Regnier-Coudert

Abstract Agent-based simulation is increasingly used in industry to model systems of interest allowing the evaluation of alternative scenarios. By this means, business managers can estimate the consequences of policy changes at low cost before implementing them in the business. However, in order to apply such models with confidence, it is necessary to validate them continuously against changing business patterns. Typically, models contain key parameters which significantly affect the overall behaviour of the system. The process of selecting such parameters is an inverse problem known as 'tuning' In this chapter, we describe the application of computational intelligence to tune the parameters of a workforce dynamics simulator. We show that the best algorithm achieves reduced tuning times as well as more accurate field workforce simulations. Since implementation, this algorithm has facilitated the use of simulation to assess the effect of changes in different business scenarios and transformation initiatives.

12.1 Introduction

Agent-based simulation is increasingly used in industry to model systems of interest allowing the evaluation of alternative scenarios. By this means, business managers can estimate the consequences of policy changes at low cost before implementing them in the business. In the case of large field workforces, agent-based simulation can provide important insight into the operational effects of new product introductions, special events and changes in demand.

T. Mamer • J. McCall (✉) • O. Regnier-Coudert
School of Computing Science and Digital Media, Robert Gordon University, Aberdeen City, UK
e-mail: j.mccall@rgu.ac.uk

S. Shakya • G. Owusu
Research and Innovation, BT Technology, Services and Operations, Martlesham Heath, UK

G. Owusu et al. (eds.), *Transforming Field and Service Operations*,
DOI 10.1007/978-3-642-44970-3_12, © Springer-Verlag Berlin Heidelberg 2013

However, in order to apply such models with confidence, it is necessary to validate them continuously against changing business patterns. Generally, models contain key parameters which significantly affect the overall behaviour of the system. Validation requires an appropriate selection of parameters such that the behaviour of the system most closely conforms to observed reality in known circumstances, typically determined from historical data sets. The process of selecting such parameters is known as 'tuning'.

Parameter tuning is what is known as an *inverse problem*, that is to say that simulation runs compared against real data are used to evaluate the simulation parameters. Candidate parameter values are set in a simulation and are evaluated in terms of the closeness of fit the output gives to historical data. Where simulations are computationally expensive, efficient search is necessary to limit the number of simulations required to select suitable parameters in reasonable time.

Workforce Dynamics Simulator (WDS) is a dynamic business simulation environment developed at British Telecom (BT). WDS enables the simulation of various working practices, organisational structures and work allocation scenarios in the light of event-driven perturbations to examine their impact on the execution of work plans based on historical or generated data. It has been used to model the BT engineering field force, simulate scenarios on service provisioning and repair and track resultant right first time measures, derived from the daily volume of jobs completed by or out of commitments. WDS has been used on a number of transformation programmes that impact upon customer experience. It has been used in developing scenarios to improve the customer appointments. WDS has also been used in assessing the impact of resourcing levels on Service Level Agreements (SLAs).

For a given scenario, WDS takes detailed input data on geographical locations, tasks to be completed and engineers available over a period of days, typically a 90-day planning period. The system simulates the performance of all tasks, including missed deadlines, rescheduled or dropped tasks, job completion times and between-job transit times. A key problem is that WDS is sensitive to choice of input parameters. These must therefore be tuned to ensure that the prediction output by WDS is as accurate as possible.

The remainder of this chapter is structured as follows. In Sect. 12.2 we describe the WDS system and explain the need for parameter tuning. Representations of parameter tuning as an optimisation problem are discussed in Sect. 12.3. Additionally a number of candidate optimisation algorithms are presented. Section 12.4 describes an evaluation methodology by which the performance of different algorithms on WDS parameter tuning is defined and fair comparisons made. Several experiments are defined, and their results are presented and analysed in Sect. 12.5. Overall conclusions are presented in Sect. 12.6.

12.2   Team Dynamics in Field Service Operations

British Telecom (BT) operates field service teams of telecommunication engineers carrying out a series of operational tasks, each within a defined geographical region. The operation of each team is complex with typically hundreds of tasks, dependencies between tasks and multiple sites requiring travel between tasks. Engineers within a geographical region are organised into multidisciplinary teams travelling in vans to task sites. Some equipment is carried as standard in vans; other equipment or parts for a task are retrieved from storage locations within the region. Periodically fluctuations in demand, changes to customer SLA and other factors necessitate reorganisations or changes to working policy for field teams. It is important therefore for management to understand the likely impact on operations of any proposed changes.

Workforce Dynamics Simulator (WDS) (Liret et al. 2009; Liret and Dorne 2009; Shah et al. 2009; Borenstein et al. 2010) is a simulation environment developed at BT which enables large-scale simulation of field service workforces and resources. WDS uses historical or generated data to investigate the execution of work plans. The system is used in BT to support business decision-making. Particular scenarios, such as predicted variation in demand or the introduction of new SLAs, can be input to the system and their likely effect on operations calculated. The results of the simulation can then be taken into account when deciding policy.

WDS takes detailed input data on geographical locations, tasks to be completed each day and engineers available on each day. The task details include among other things their type and location, what skills are required to complete these tasks and details of any existing appointments. The engineer details include among other things their rosters, skills, preferred working areas and starting locations.

WDS uses this input data to simulate the completion of tasks over a number of days. The process starts with the time at which the task is reported and enters into the system and continues with the tasks being allocated to an engineer which then travels to the task's location and completes the task (or fails to do so). The allocation of tasks to engineers is done through a scheduler which uses a rule-based allocation system to decide which task is allocated to an available engineer. This scheduler takes into account geographical domains and locations, technical skills that are available and required, importance of tasks as well as their appointments and committed times.

After a simulation is run, WDS provides a detailed report, containing daily expected numbers on productivity, volume of jobs completed, volume of jobs failed, volume of jobs carried over to the next day and the percentage of successes. This can then be used to support decision-making in the business. See Fig. 12.1 that shows a dataflow diagram of WDS.

In this chapter, we are concerned with ensuring the validity of WDS so that simulations are as reliable as possible. BT maintains data sets of actual operational outcomes for a variety of geographical locations over 30–90 daytime periods. WDS

Fig. 12.1 WDS dataflow

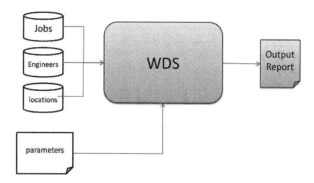

can be run on actual task lists for these periods and the output report compared with known outcomes.

It is observed that the predictive accuracy of WDS is highly sensitive to five key simulation parameters. These are:

- Average speed (AS): the average speed at which the engineers travel between task sites.
- Inefficiency factor (IF): an estimated number of minutes for which engineers are idle after each job completion. No jobs can be assigned to them during this time.
- Overtime allowed (OA): the additional time that engineers spend completing a task, after the rostered time has elapsed. This affects completion rates.
- Lateness allowed (LA): the amount of time engineers can use to complete a task when they have arrived on location too late.
- Distribution of reassigned tasks (RT): the proportion of jobs which have a strong probability of failure after the first assignment. These tasks will then be reassigned.

Alterations to these parameters often yield a significant change in the output of WDS simulations. The challenge then is to tune these WDS parameters so that the simulation of test scenarios yields results as close as possible to the known outcomes. Once the parameters are tuned to a test scenario, WDS can be used to simulate the effect of changes in other setups such as the effect of adding more technicians to the workforce or adding more jobs to the work stack.

A naïve approach to WDS tuning is the *parameter sweep*: a large grid of combinations of the possible values for these tuning variables is simulated and evaluated exhaustively. More formally, ranges are defined for each of the five tuning variables AS, IF, OA, LA and RT as well as a *step-size k*. The step-size defines how many values in between these ranges will be tested. Given a step-size of k, then for any particular parameter v with range v_{min} to v_{max}, k equally spaced parameter values will be tested, including v_{min} and v_{max}. The *ith* value in this sequence is denoted v_i and is calculated from the formula $v_i = v_0 + (i - 1)(v_{max} - v_{min})/(k - 1)$. Step-size k may of course vary for each parameter v. We denote by variable I a vector of indices $(i_1, i_2, i_3, i_4, i_5)$ where each i_j is in the range 1 to k_j. We define the parameter set indexed by I, $p(I)$, to be the vector of parameter values

Fig. 12.2 Parameter Sweep
Algorithm

```
I = (1,1,1,1,1)
p = p(I)
p* = p
while I < (k₁,k₂,k₃,k₄,k₅) {
        I++
        if eval (p(I)) > eval (p*) {
                p* = p(I)
        }

}
return p*, eval (p*)
```

$(AS\_i_1, IF\_i_2, OA\_i_3, LA\_i_4, RT\_i_5)$. We also define a functional notation to represent the process of inputting a vector p of parameters to the simulation. The notation *eval (p)* is used to represent a numerical measure of that output against known outcomes. More detail on the nature of eval is given later, but for now we assume we are maximising it. As a convenient notation for cycling through the indices, we write I++ to mean a milometer-style increment to the vector. That is, each time ++ operates on *I*, the following two rules are applied:

- The rightmost index is incremented by 1.
- If any index j is at its maximum value k_j, and it is to be incremented, then it is set back to 1, and the next index to the left is then incremented by 1.

We note that this increment operator also imposes an ordering on the values of *I*.

With this notation in place, the Parameter Sweep Algorithm is as follows (Fig. 12.2):

The algorithm starts with the minimum value in each parameter range and systematically iterates through all combinations of parameter values. For each parameter set p, *eval (p)* is computed. Each computation of *eval* requires a simulation run. This quickly becomes computationally expensive as greater precision on parameter values is sought. We note that Parameter Sweep has a computational complexity of $O(k_1 k_2 k_3 k_4 k_5)$.

Typical value ranges for the WDS parameters are given in Table 12.1 in the following section. In this example, there are $32 \times 64 \times 64 \times 16 \times 16 = 33{,}554{,}432$ parameter combinations. As each parameter set needs to be simulated, this is the minimum number of times that the simulation would have to be run using the Parameter Sweep Algorithm on typical value ranges.

To run one simulation of one week (7 days) takes on average 105 s.[1] So if we were to test each of these possible combinations for 1-week simulations, it would take around 3,500,000,000 s, which would amount to almost 113 years. However, WDS usually simulates over 90 days which makes the time intractable.

A further limitation of Parameter Sweep is that no value between the steps will be tested. In a complex and irregular simulation such as this, it is unlikely that optimal parameters will be found at points on an evenly spaced grid of values. It is

---

[1] This time was measured on an Intel(R) Core(TM) i5 CPU M 520 @ 2.40 GHZ, 1.86 GB of RAM.

Table 12.1 The five decision variables, their ranges and the number of bits allocated to them

	Average speed	Overtime allowed	Lateness allowed	Inefficiency factor	Reassigned tasks dist.
Minimum	10	0	60	2	2
Maximum	41	63	123	17	17
No. of values	32	64	64	16	16
Bit required	5	6	6	4	4

clear therefore that a naïve parameter sweep is not an appropriate method to find the best set of parameters. A more efficient algorithm is required.

12.3 Representations and Algorithms

In this section, we will investigate algorithms that search the ranges of possible parameter values in order to find the best parameter set possible with the computational resource available. Traditional gradient-based approaches are infeasible here due to the irregularity of the search space, so we turn instead to metaheuristic algorithms. The general workflow is diagrammed in Fig. 12.3. A metaheuristic algorithm maintains a population of solutions to a problem, normally initialised at random. At each iteration, solutions in the population are evaluated, and this information is used to determine a successor population of solutions to be examined. The process by which the successor solutions are produced is called the metaheuristic. The process continues until some termination criteria have been reached, for example, the available computational resource has been exhausted.

A wide variety of metaheuristics exist, and they have been successfully applied to a broad range of problems in science and engineering. There are two essential requirements for application of a metaheuristic to a problem. First, solutions need to be represented in a way that can be manipulated by the metaheuristic. Second a procedure must exist by which any selected solution can be evaluated. Here solutions will represent value choices for the five WDS parameters, and the evaluation will be based on comparing outputs from the simulation with known outcomes from historical operations. Of particular interest will be to find high-quality parameter sets with a restricted number of solution evaluations. In the rest of this section, we define some representations of parameter sets, and we select a set of algorithms to try on the problem. We define the *eval* function precisely in Sect. 12.4.

Fig. 12.3 General
workflow of a metaheuristic
algorithm

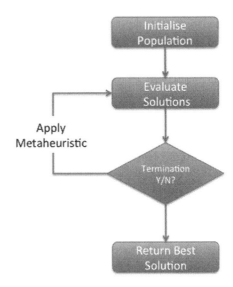

12.3.1 Representations

The natural way to represent WDS parameters is as a vector of five floating-point
numbers. We call this the *real-valued encoding*, and it will be used directly by some
of the algorithms that we study. Other algorithms in this work require a *binary-
valued encoding*. This is defined as a single binary string that can then be decoded to
recover values for each variable. There is a design choice to be made as to how long
the binary string is and how many bits are allocated to each variable. After some
experimentation and examination of typical variable ranges, we adopted the scheme
displayed in Table 12.1.

Thus, each solution generated by an algorithm using the binary encoding
consists of 25 bits and is converted into a set of five parameters by binary to integer
conversion of those sections of the string devoted to each parameter. For example,
the first five bits are converted into a natural number between 0 and 31. Then the
minimum value for AS, $min(AS) = 10$, is added, which gives us a number between
10 and 41, corresponding to the range given in Table 12.1. This process is then
replicated mutatis mutandis for OA, LA, IF and RT, respectively.

12.3.2 Algorithms

In this study, we perform comparative experiments with a range of metaheuristic
search algorithms. They include population-based Evolutionary Algorithms
(EA) (Goldberg 1989; Davis 1996), Estimation of Distribution Algorithms (EDA)

(Muehlenbein and Paaß 1996; Larrañaga and Lozano 2002; Shakya and McCall 2007) as well as one non-population-based algorithm.

All of these algorithms conform to the workflow given in Fig. 12.3 but differ in the specific metaheuristic used to generate successor solutions. We set out the algorithms and explain the metaheuristic below. Further details of the algorithms may be found in the references given.

- **Genetic algorithm (GA)** (Goldberg 1989; Mitchell 1997), an EA. Its metaheuristic generates successor populations of solutions using operators inspired by natural evolution: selection, crossover and mutation. We have used both binary encoding and real encoding to represent solutions for GA.
- **Population-based incremental learning (PBIL)** (Baluja 1994; Baluja 1995), an EDA. The PBIL metaheuristic estimates an independent probability distribution for each solution variable from a set of selected high-value solutions. This distribution is then sampled to generate successor populations of solutions. We have used a binary encoding to represent solutions for PBIL.
- **Distribution estimation using Markov random field with direct sampling (DEUM$_d$)** (Shakya 2006; Petrovski et al. 2006; Shakya et al. 2010), also an EDA. The DEUM metaheuristic estimates the distribution of value across solutions and then samples higher-quality solutions with a high probability. DEUM$_d$ also uses a binary encoding.
- **Univariate marginal distribution algorithm for continuous domains (UMDAc)** (Gonzalez et al. 2002; Yuan and Gallagher 2009), an EDA. The UMDAc metaheuristic operates on a floating-point encoding and estimates a Gaussian distribution of solution variable values from a set of selected high-value solutions.
- **Simulated annealing (SA)** (Kirkpatrick et al. 1983; Otten and van Ginneken 1989). SA iterates on a sequence of single solutions, which for our purposes we regard as a population of size 1. The SA metaheuristic is an energy-minimisation technique modelled on the cooling properties of metals. We have used both binary and real encodings for SA.

12.4 Evaluation

In this section we use the simulation and given historical outcome data to provide the *eval* function for solutions. Each algorithm defined in Sect. 12.3 will be implemented as a *wrapper* for the simulation: that is, a call on *eval (p)* for a given solution *p* will cause the parameters defined by *p* to be passed to WDS. WDS will then simulate a given historical scenario using these parameters. The output of WDS will make a set of comparisons of the simulation output with observed outcomes and combine these into a single value which will be returned as the value of *eval (p)*. The wrapper schema is diagrammed in Fig. 12.4.

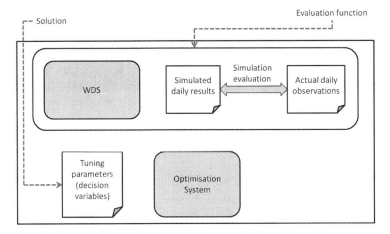

Fig. 12.4 Wrapper schema for WDS

This *eval* function is composed of a Pearson Correlation Score (Rodgers and Nicewander 1988) (measuring the correlation between two variables overtime) and a Mean Absolute Percentage Error (MAPE) (Makridakis and Hibon 2000) (measuring accuracy in a fitted time series) calculated between the daily simulated outcomes and observed outcomes over a period of d days.

The observed outcomes used in this work are comprised of real world observations on tasks belonging to two different categories: *Provision* and *Repair*. Provision tasks deal with proactively installing some telecommunications infrastructure needed for future services, whereas Repair tasks are reactively carried out in response to customer's immediate needs.

For each day i, the observed outcomes consist of the following values:

- **Provision Performance** (PP_i): the percentage of Provision tasks successfully performed on day i
- **Repair Performance** (PR_i): the percentage of Repair tasks successfully performed on day i
- **Provision Failures** (PF_i): the number of Provision tasks which has failed on day i, that is completed but after the scheduled time
- **Provision Successes** (PS_i): the number of Provision tasks which was completed successfully on day i
- **Provision Totals** (PT_i): the total number of Provision tasks which has been completed on day $i (PT_i = PF_i + PS_i)$
- **Repair Failures** (RF_i): the number of Repair tasks which has failed on day i (completed but after its committed time)
- **Repair Successes** (RS_i): the number of Repair tasks which was completed successfully on day i that is completed but after the scheduled time
- **Repair Totals** (RT_i): the total number of Repair tasks which has been completed on day $i (RT_i = RF_i + RS_i)$

12.4.1 Pearson Correlation Score

The WDS outputs each of these Provision and Repair measures for each day, and so we can make comparisons of simulated outcomes with observed ones. First, for each simulated outcome, x, we compute the Pearson correlation coefficient against the observed outcomes, $r(x)$, as follows:

$$r(x) = \frac{\sum_{i=1}^{d} \left(xAct_i - \overline{xAct}\right)\left(xSim_i - \overline{xSim}\right)}{\sqrt{\sum_{i=1}^{d}\left(xAct_i - \overline{xAct}\right)^2}\sqrt{\sum_{i=1}^{d}\left(xSim_i - \overline{xSim}\right)^2}} \tag{12.1}$$

where $xSim$ is the simulated outcome for day i and $xAct$ is the actual observed outcome for day i.

Then we combine the Pearson correlations for all eight values together into a weighted Overall Simulation Pearson Score which we attribute to the input parameters p, $OSPS\,(p)$:

$$OSPS(p) = \left(\alpha \times \frac{|r(PP)| + |r(PR)|}{2}\right)$$
$$+ \left((1-\alpha) \times \frac{|r(PS)| + |r(PF)| + |r(PT)| + |r(RS)| + |r(RF)| + |r(RT)|}{6}\right) \tag{12.2}$$

where correlation weight α is a parameter provided to the function in order to balance the weight of the performance percentages (PP and PR) and the task counts. In this work, $\alpha = 0.7$, assigning more weight to performance percentages than to the counts of tasks.

12.4.2 Mean Absolute Percentage Error

Second, for each simulated outcome x, we calculate the Mean Absolute Percentage Error (MAPE) as $\varepsilon(x)$ over the period of d days:

$$\varepsilon(x) = \frac{1}{d}\sum_{i=1}^{d}\left|\frac{xSim - xAct}{xSim}\right| \tag{12.3}$$

Then we combine the MAPE scores for all eight values together into one Overall Simulation MAPE Score, $MAPE\,(p)$ which we attribute to the input simulation parameters p:

$$MAPE(p) = \left(\beta \times \frac{|\varepsilon(PP)| + |\varepsilon(PR)|}{2} \right)$$
$$+ \left((1-\beta) \times \frac{|\varepsilon(PS)| + |\varepsilon(PF)| + |\varepsilon(PT)| + |\varepsilon(RS)| + |\varepsilon(RF)| + |\varepsilon(RT)|}{6} \right)$$

$$(12.4)$$

where error weight β, similar to α above, is a parameter provided to the function in order to balance the weight of the performance percentages (PP and PR) and the values counting tasks. In this work, $\beta = 0.7$.

12.4.3   Overall Simulation Score

Finally, $OSPS(p)$ and $MAPE(p)$ are combined to compute one overall Simulation Score $simScore(p)$ for each set of input parameters p:

$$simScore(p) = \left(\gamma \times OSPS(p) + ((1 - \gamma) \times (1 - MAPE(p))) \right) \qquad (12.5)$$

where γ allows the balancing between $OSPS(p)$ and MAPE(p). In this work, we set $\gamma = 0.5$, assigning equal weight to both scores. As this final score aims to maximise the correlation while minimising the errors, $(1 - MAPE(p))$ was used. The parameter set resulting in the highest overall simulation score is considered to be the best. The wrapper schema can now use a number of different algorithms to find *solutions* which consist of (or can be converted to) sets of tuning variables p and use WDS to calculate their evaluation:

$$eval(p) = simScore(p) \qquad (12.6)$$

The optimisation problem is to find the set of variables V that maximise $eval(p)$.

12.4.4   Objectives

To evaluate the results achieved, we compared the different algorithms not only with each other, but we also used a benchmark which we aimed to outperform. This benchmark is the performance of the *Parameter Sweep* algorithm, restricted to a computationally tractable number of evaluations, that is to say using a very coarse-grained grid. We ran Parameter Sweep using the following step-sizes and with standard ranges set for all the metaheuristic algorithms. These are shown in Table 12.2.

Using this setup, the tool ran $4 \times 3 \times 7 \times 8 \times 8 = 5,376$ simulations. The best result found with Parameter Sweep had an *eval* of 0.8077 (see 12.6). Note that we would

Table 12.2 Parameter Sweep benchmark configuration

	Min	Max	Step-size	No. of values
AS	10	40	10	4
OA	0	60	30	3
LA	60	120	10	7
IF	2	16	2	8
RT	2	16	2	8

expect this to be a poor result given the coarse-grain nature of the grid and the known sensitivity of WDS to parameter values. During our experiments, our objective was to address the two following challenges:

1. Surpass this benchmark (e.g. consistently find a set of tuning variables which result in a better fitness than Parameter Sweep).
2. Find results at least as good as the benchmark but running significantly fewer simulations.

12.5   Experiments and Results

12.5.1   Algorithm-Specific Settings

We ran each of the presented algorithms on our WDS tuning problem using a number of different algorithm settings. Because simulations in WDS are very time-consuming to run, most algorithms ran for almost a whole day before results could be analysed. Some even ran over several days (e.g. DEUM$_d$ and SA using real encoding). Because of these long execution times, we were not able to run every possible algorithm configuration multiple times extensively. Instead, we ran our experiments in two stages. In the first stage, we tested a wide range of configurations for each algorithm by running each of them twice. Based on the results achieved from those, we selected the most promising configuration for each algorithm. In the second stage, we ran the chosen configurations six times.

For all algorithms except SA, which operates on a single solution per iteration, we set a population size of 50 solutions. We also allowed for the two best solutions of each generation to pass over to the next generation unaltered. As it was our aim to outperform our benchmark, we did not allow any of our algorithms to run more evaluations than our benchmark (5,376). To determine an appropriate maximum number of generations, we investigated the point at which each algorithm converged in our first stage experiments. We consequently ran GA and UMDAc for a maximum of 50 generations, PBIL for 60 generations and DEUM$_d$ for 100 generations. We allowed SA to execute a maximum of 3,000 generations. In light of this relatively small amount of evaluations, we set an accordingly large SA cooling rate of 0.05 in both real- and bit-encoding experiments. We set PBIL to use a selection size of 20 and a learning rate of 0.3. We used GA with one-point crossover, a crossover probability of 0.7 and a tournament selection operator. Our first stage experiments indicated that using a real encoding, GAs find the best solutions using a mutation rate of 0.1, while a mutation rate of 0.05 worked best when using a bit encoding. For DEUM$_d$, cooling rate was set to 6.0. We refer the reader to referenced literature for more details on these algorithm configurations.

12.5.2   Solution Caching

Often, metaheuristic algorithms will generate solutions that have been tested for fitness previously in the same execution, especially when the search converges into one area of the search space. As our WDS tuning objective function is so expensive to run, we avoided multiple simulations using identical decision variables. We implemented a *Solution Cache*, a table in the working memory, storing each solution which has already been evaluated in the current execution and its *eval*. Each time a solution is evaluated, it is checked whether it is already stored in the Solution Cache. If it is, the previously calculated evaluation is reused avoiding a rerun of the simulation.

12.5.3   Results

The evaluation mean and standard deviation (st Dev), as well as the evaluation score of the overall best solution found, are given in Table 12.3. It also gives the mean improvement on the benchmark for the best solutions found for each algorithm. Finally, Table 12.3 also gives the mean number of evaluations (i.e. simulations) run by each algorithm. This value is different from the maximum number of allowed evaluations due to solution caching. Figure 12.5 plots the mean best and best overall evaluation achieved for each algorithm. For comparison, we also plot the benchmark evaluations in Fig. 12.5.

12.5.4   Discussion

In this work, we set two different challenges. The first was to find better solutions that can be found by a benchmark Parameter Sweep using a feasible number of simulations. This challenge was met by all of the tested algorithms.

The second challenge was to maintain solution quality but using significantly fewer evaluations. This challenge was also met by all of the tested algorithms. In fact, it was clear from the first stage of our experiments that such a high number of simulations are never required to find good results, irrespective of the algorithm configurations tested. For our second stage of experiments, we therefore set the

Table 12.3 Results

Algorithm	Mean	StDev	Best Eval	Improvement on Benchmark		Mean Number of Evaluations
				Mean	Best	
UMDAc (real)	**0.8312**	0.0054	**0.8389**	**0.0235**	**0.0312**	1462
PBIL (bit)	0.8209	0.0028	0.8252	0.0132	0.0175	1283
GA (real)	0.8226	0.0053	0.8284	0.0149	0.0207	1090
GA (bit)	0.8140	0.0058	0.8230	0.0063	0.0153	**334**
SA (real)	0.8240	0.0036	0.8293	0.0147	0.0179	2733
SA (bit)	0.8232	0.0054	0.8321	0.0155	0.0244	1410
DEUMd (bit)	0.8236	0.0025	0.8279	0.0163	0.0216	2926

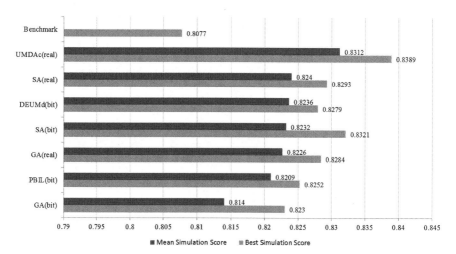

Fig. 12.5 Mean evaluation and maximum evaluation achieved with each algorithm. They are arranged after the benchmark in descending order with respect to the mean fitness

population size and *maximum generation* parameters so that more than 3,000 simulations are unlikely to be run.

From the results presented in Fig. 12.3, we can clearly see that UMDAc performs best on the WDS tuning problem, both in terms of the overall best result found in six executions and in each individual execution. UMDAc needs to run around 1,500 simulations, which is a significant improvement compared to the benchmark of 5,376; however, some algorithms, notably GA, can exceed the benchmark with far fewer simulations, though with some penalty in solution quality. DEUM$_d$, a close third in terms of evaluation quality, has the lowest variation, therefore finding good solutions reliably. However, this is at the expense of a comparatively large number of simulations during its execution. SA (bit), SA (real), PBIL and GA (real) behave comparably in terms of the fitness of solutions found; however, SA (real) has to run a much larger number of simulations during its execution. Finally, GA (bit) is the worst in terms of the fitness of the solutions found; however, in most cases it is still finding solutions which are better than the benchmark. It is also worth pointing out that it runs a remarkably small amount of simulations during its execution, less than half the amount of the next lowest algorithm we tested. This is due to convergence on suboptimal solutions with significant use of solution caching. GA (real) has similar properties, using less than a third of the evaluations used by UMDAc and less than a tenth of those used by the Parameter Sweep.

The results suggest a trade-off between finding the best overall solution and the number of evaluations required to get there. Considering that WDS simulations take a long time to run, using an algorithm which converges more quickly may be worth a small penalty in solution quality.

The results also suggest that better solutions are found when they are encoded as real numbers, the natural encoding for this problem. In particular, for those algorithms (GA and SA), which offer a choice of both encoding schemes, the real encoding always delivers better solutions compared to the bit encoding.

12.6   Conclusions

WDS is an important tool for supporting strategic decision-making at BT. Its value is in the ability to make reliable forecasts of the likely effect on operations of policy and demand changes, and therefore this value rests on valid and reliable simulation, achieved through precise tuning of five key input parameters. We have shown in this work that metaheuristic algorithms can be used effectively and efficiently to automate the tuning of WDS parameters. Our experiments showed that metaheuristics find better tuning variables than a naïve Parameter Sweep in a significantly shorter period of time.

The algorithms show variation in performance depending on the encoding used, and, within the overall speed-up obtained by all algorithms over Parameter Sweep, there is some trade-off between quality of solution and the number of simulation runs needed to attain it.

In the wider context of management decision-making, the broad value of the metaheuristics is that they systematically select parameters that have experimental validity in a reasonable timescale, and, from that point of view, the metaheuristic algorithms each add similar value. In fact, all of them are now incorporated for standard use in the WDS tool.

References

Baluja S (1994) Population-based incremental learning: a method for integrating genetic search based function optimization and competitive learning. DTIC, Fort Belvoir

Baluja S (1995) An empirical comparison of seven iterative and evolutionary function optimiza-
    tion heuristics. Carnegie Mellon University, Pittsburgh
Borenstein Y, Shah N, Tsang EPK, Dorne R, Alsheddy A, Voudouris C (2010) On the partitioning
    of dynamic workforce scheduling problems. J Schedul 13:411–425
Davis T (1996) The handbook of genetic algorithms. Van Nostrand Reinhold, New York
Goldberg DE (1989) Genetic algorithms in search, optimization and machine learning. Addison-
    Wesley, Boston, MA
Gonzalez C, Lozano JA, Larrañaga P (2002) Mathematical modelling of UMDAc algorithm with
    tournament selection. Behaviour on linear and quadratic functions. Int J Approx Reason
    31:313–340
Kirkpatrick S, Gelatt CD, Vecchi MP (1983) Optimization by simulated annealing. Science
    220:671–680
Larrañaga P, Lozano JA (eds) (2002) Estimation of distribution algorithms: a new tool for
    evolutionary computation. Springer, Berlin
Liret A, Dorne R (2009) Work allocation and scheduling. In: Voudouris C, Owusu G, Dorne R,
    Lesaint D (eds) Service chain management: technology innovation for the service business.
    Springer, Berlin
Liret A, Shepherdson J, Borenstein Y, Voudouris C, Tsang E (2009) Workforce dynamics
    simulator in service operations scheduling systems. In: Conference on enterprise information
    systems
Makridakis S, Hibon M (2000) The m3-competition: results, conclusions and implications. Int J
    Forecast 16(4):451–476
Mitchell M (1997) An introduction to genetic algorithms. MIT Press, Cambridge
Muehlenbein H, Paaß G (1996) From recombination of genes to the estimation of distributions
    I. Binary parameters. In: Parallel problem solving from nature PPSN IV. pp 178–187
Otten RHJM, van Ginneken LPPP (1989) The annealing algorithm. Kluwer, Boston, MA
Petrovski A, Shakya S, McCall J (2006) Optimising cancer chemotherapy using an estimation of
    distribution algorithm and genetic algorithms. In: Genetic and evolutionary computation
    conference GECCO. pp 413–418
Rodgers JL, Nicewander WA (1988) Thirteen ways to look at the correlation coefficient. Am
    Statistic 42(1):59–66
Shah N, Tsang EPK, Borenstein Y, Dorne R, Liret A, Voudouris C (2009) Intelligent agent based
    workforce empowerment. In: KES-AMSTA'09. pp 163–172
Shakya S (2006) DEUM: a framework for an estimation of distribution algorithm based on Markov
    random fields. The Robert Gordon University, Aberdeen
Shakya S, McCall J (2007) Optimization by estimation of distribution with DEUM framework
    based on Markov random fields. Int J Automat Comput 4:262–272
Shakya S, Brownlee A, McCall J, Fournier F, Owusu G (2010) DEUM: a fully multivariate EDA
    based on Markov networks. In: Chen Y (ed) Exploitation of linkage learning in evolutionary
    algorithms. Springer, Berlin
Yuan B, Gallagher M (2009) Convergence analysis of UMDAC with finite populations: a case
    study on flat landscapes. In: Genetic and evolutionary computation conference GECCO. pp
    477–482

Chapter 13
Effective Engagement of Field Service Teams

Tanya Alcock and Jonathan Malpass

Abstract Successfully engaging and motivating any work force is vital; how to accomplish such a challenge is the focus of this chapter. It is the shared view of the authors that attaining employee engagement requires a finely tuned mix of actions based upon understanding the workforce from a qualitative and quantitative viewpoint. The qualitative view, based on social science research, reveals insights into peoples' actions, their behaviours, attitudes and values. The quantitative view provides the essential knowledge and information to facilitate the employee's job in hand as well as enable the workforce to reach their full potential. This chapter outlines nine components which emerged from this approach, components that any organisation needs to understand and action to influence employee engagement.

13.1 Introduction

People are the key asset of any organisation, and those deploying field-based workforce teams cannot afford to be indifferent or ill informed about their workforce. If they are, any efforts an organisation makes on forecasting, modelling and delivery of field operations will be to limited avail. One thing is clear: an engaged and motivated workforce is capable of achievements well beyond simply boosting productivity (i.e. working harder and faster). Engaged employees are a rich source of recommendations for improvements to current (inefficient) working practices, they can provide innovative new business ideas, they can pull together to produce fantastic feats during times of crisis, and they can win over new customers as well as retain defecting customers, all of which contribute to the performance of the organisation as well as enhancing its competitive advantage.

Furthermore, engagement is a two-way street between the employee and the organisation, with both parties benefiting when it is achieved and sustained.

T. Alcock (✉) • J. Malpass
Research and Innovation, BT Technology, Services and Operations, Martlesham Heath, UK

G. Owusu et al. (eds.), *Transforming Field and Service Operations*,
DOI 10.1007/978-3-642-44970-3_13, © Springer-Verlag Berlin Heidelberg 2013

Successful employee engagement goes beyond just benefiting the business; employee engagement can also be the key to transforming the working lives of the employees themselves. While an engaged workforce is a happier, well-balanced and focused team, engagement also works for individuals, allowing people to have a greater sense of well-being, understand how they contribute to their employer's successes, gain emotional buy-in to their work and potentially banish those 'bad day at work' blues. Individuals who are intrinsically motivated, as well as enabled, are willing to give their best.

This chapter describes a fresh perspective of employee engagement, identified through the authors' research. By using a powerful combination of qualitative and quantitative research techniques, and distilling those findings, nine key components have emerged that organisations need to understand and action to influence employee engagement. These components are management style; interpersonal relationships and team collaboration; incentives; measures and targets; knowledge and relevant data; continuous improvement; job design; career, mastery and advancement; and work environment. All the different components addressed are essential requirements to be considered by organisations when engaging and motivating a workforce. In short, it is an inclusive list of essential components for any organisation rather than a pick and mix approach.

13.2 Why Bother with Engagement?

The results of successfully engaging and motivating a workforce are evident. Tamkin et al. (2008) report a 10 % increase in employee engagement can potentially increase profits by £1,500 per employee per year. Sears, the US retailer, carried out their own investigation into employee engagement which resulted in the development of the employee-customer-profit chain. They discovered that a 5-unit enhancement into employee attitude (aligned to engagement) drives through an increase in customer impression which is worth 0.5 % increase in revenue growth (Rucci et al. 1998). Harter et al. (2002) found that businesses which typically scored higher for employee engagement experienced higher profitability; they also found that productivity via revenue/sales was on average $80,000–$120,000 higher per month (with one organisation the difference was more than $300,000).

If engaged employees can benefit companies, disengaged employees have an associated cost all of their own. The disengaged are more likely to leave companies resulting in raised recruitment costs, as well take more sick days than their engaged colleagues, on average 6.19 days off a year compared to an average of 2.69 sick days per year (MacLeod and Clarke 2011 and references therein). As sickness absence cost the UK economy in excess of £17bn in 2010 (CBI Report 2011), engaging your employees has a lot of potential benefits.

13.3 Background and Approach

It is the authors' opinion that attaining high levels of employee engagement requires a mix of activities based upon understanding the workforce from both a qualitative and quantitative viewpoint. The tactic of using a joint approach has arisen from more than 30 years combined experience, including understanding and working with large field-based engineering communities within a corporate environment. In addition to working with field-based engineers, the authors' experience extends to workforces within more traditional office-based locations, including knowledge workers, contact centres, sales desks and data centres within small- and medium-sized business operations as well as consumer operational support.

The qualitative research techniques and methodologies used include ethnography (O'Reilly 2004), open question interviewing and grounded theory (Strauss and Corbin 1994). These have afforded a detailed understanding of workforces and the drivers behind employees' behaviours. Insights gleaned from hundreds of hours spent shadowing and observing workers as they carry out their day-to-day work, watching what they actually do, who they interact with and the barriers they face while completing their job, are essential to establish a rich qualitative picture (often combined with interviews and questionnaires). It is these techniques based in social sciences that have contributed to the overall understanding of employee behaviour—what people do and why—and what ultimately drives engagement.

A complement to the rich behavioural picture has been the use of techniques more typically associated with Lean Methodologies (Womack and Jones 2003) and Six Sigma (Pande et al. 2000). Research techniques such as root cause analysis (Wilson et al. 1993), statistical process control (Oakland 1996) and randomised controlled trials (Haynes et al. 2012) not only complement the behavioural studies but also yield the 'hard' (quantifiable) data, an essential requirement of any business environment. It is typically the hard data which will sway senior management decisions; by coupling hard numerical data with behavioural evidence, giving additional weight and insight which can be absent from the numbers alone, ensures senior management are enabled to make more informed decisions.

13.4 What Is Engagement?

During this chapter the authors use the term engagement as a generic term referring to both the attitude and behaviours of employees, in particular the blend of positive attitudes and resulting behaviours (and actions) that ensue when an individual is actively engaged in their work and with their organisation. This includes how happy workers feel, how positively they look upon their company, how well they regard their co-workers and how involved they are with the work they are undertaking. It also includes the employee's willingness to put in extra effort (aka 'goes the extra mile') and provide additional input into a job and the organisation. Engagement also includes how intrinsically motivated the individual is by their work, and therefore motivation is considered as a fundamental part of the engagement story. Job satisfaction and morale, common terms used in the literature and by organisations, are also considered to be vital subsets of engagement.

There is a wealth of literature on employee engagement and motivation theories and how people should be managed and led (e.g. McGregor 1960; Herzberg et al. 1959; Nohria et al. 2008; MacLeod and Clarke 2011). The authors take the stance that humans are self- or intrinsically motivated to work as part of their natural desire to grow, be part of something important and succeed as individuals (e.g. Ryan and Deci 2000; Pink 2009). You cannot enforce motivation or make people be motivated: it comes from within. Along with the assumption that, in general, employees are naturally motivated, the authors consider that motivating employees is about creating the right environment or culture within which people can flourish. Therefore engagement is as much about actively removing barriers or inhibitors which prevent employees from getting on and doing their job as it is about taking action, because when it comes to motivation and engagement, it is all too easy to destroy what is often naturally present.

13.5   The Key Components of Employee Engagement

This section describes the nine components of employee engagement that the authors, through their experience and distinctive approach, have deemed to be vital to employee engagement. A wide range of subjects is covered, which can appear quite daunting, but all the components addressed are essential requirements.

13.5.1   Management Style

Our experience has led us to realise that perhaps the greatest influence on any employee is their immediate manager. How that manager chooses or believes to manage is of fundamental importance. A simple conversation with any manager about the responsibilities and practicalities of their role will often reveal their attitudes and beliefs about their people and uncover their management style and ethos.

The best people managers instinctively understand the potential within their employees: they are capable, driven and self-motivated individuals rather than the 'inert' employees who can only be cajoled into action via extrinsic rewards (first challenged by McGregor in his Theory X vs. Theory Y work (1960). They understand that trust and belief in their employees are essential and that their key role is one of an enabler. Their job is to coach and to remove the barriers that prevent employees from doing their job. They give employees autonomy and successfully empower them by giving them appropriate and necessary levels of information, responsibility and control. Empowering managers develop their employees' abilities to make decisions. They instinctively know that the imagination, creativity and ingenuity of employees can be used to solve work problems. In an effort to develop and support a culture of empowerment, managers facilitate and direct information and workflow throughout the organisation. They serve as leaders and role models for others through their efforts (Brower 1995). Good managers instinctively give their employees autonomy so they can self-direct (Pink 2009) and they act as both a catalyst and a nourisher to their employees (Amabile and Kramer 2011).

In contrast, poor people managers often place a strong emphasis on the need to observe and control their employees; they manage people with the aim of achieving compliance rather than engagement. Such managers often place high importance on measures and may even be proud of their ability to 'manage' the measures rather than having pride in their employees' abilities and achievements. These behaviours are driven by misguided assumptions about what intrinsically motivates employees (Deci 1975). They wrongly assume that employees dislike working, need to be directed, dislike responsibility and must be controlled and threatened before they will work hard enough (McGregor 1960). They assume that the average person is only motivated by rewards, incentives and avoidance of penalties (see Sect. 13.3). Poor managers are, intentionally or unintentionally, toxins to creativity and inhibitors to their employees' true potential (Amabile and Kramer 2011).

13.5.2   *Interpersonal Relationships and Team Collaboration*

Interpersonal relationships within the workplace stretch beyond management. Individuals and their levels of engagement and morale are strongly influenced by other relationships, such as those with colleagues or peers within teams and other groups with who they have routine contact. Many individuals enjoy the opportunity to work with others on a regular basis, often referred to as 'teamwork' and 'team spirit'.

However, in our experience the interpersonal relationships which influence individuals can often go beyond organisational boundaries. Within customer-facing teams these relationships often include the customers themselves, even including the customer's customer. Interacting with, and helping, a customer is one of the key drivers of job satisfaction and morale seen in both contact centres and field teams. The level of 'emotional' commitment amongst employees towards the customers with whom they interact can be a powerful asset which should be utilised.

Given that some of the interactions with customers can be transient, this strong sense of commitment to customers should be commended, even encouraged, because of the benefits to the organisation, the employee and the customer. Successful emotional engagement with customers is one of the prerequisites to providing a superb customer interaction and experience (Shaw and Ivens 2002; Reichheld 2000).

Interpersonal relationships are both formal and informal and usually reciprocated. An engineer may have an appointed coach to whom they can go to for help, but they will also have a widespread network of informal contacts to which they can turn if specific advice or local knowledge is needed. Typically, such knowledge is not formally captured by knowledge management systems but relayed between individuals during early morning meetings in the office or a quick phone call—it is fluid, fast and specific.

The sharing of such advice and knowledge through informal networks goes beyond the simple exchange of information required to do the job. It also is an opportunity to reinforce bonds which have been formed within workplace teams and enhance team spirit. This can be increasingly important to morale and engagement if team get-togethers are limited due for logistical or geographical reasons. Such opportunities should be encouraged and facilitated because they benefit both the company (through knowledge sharing) and employees (through enhanced camaraderie). A sense of belonging and cooperation are important boosters for morale within any workplace. Hardy's PhD study on morale (2010) found that interpersonal relationships were one of three main contributors to positive morale (along with future/goal and emotional aspects). Constructive, supportive relationships are associated with high morale; divided, isolated relationships or a bad atmosphere is linked to low morale. Hardy (2010) also considered morale to be a contagion, so high morale can be spread from individual to individual, indicting the importance of encouraging opportunities for team interactions.

13.5.3   Incentives (and the Associated Stick)

'Pay is not a motivator' (W. Edwards Deming)

Incentives are the subject of much debate within both social science and management science. While it has long since been established that humans are motivated by matters beyond simple financial rewards (e.g. Maslow 1943; Herzberg 1968; Pink 2009), the belief that employees are only motivated by rewards or by the fear of external punishment is endemic within many large organisations. Deci's studies (1975; Deci and Ryan 1985) into intrinsic motivation revealed a different view of what made humans 'tick'. Herzberg (1968) drew the distinction between 'motivation' and 'movement'; movement is short term and often via a 'stick' or 'KITA' (kick in the a**), whereas motivation is often long term and driven by the individual's intrinsic motivation to achieve, grow and learn. In short, rewards and incentives don't create lasting commitment or motivate or engage employees; they just cause people to temporarily comply with what is required to earn the reward (or avoid the stick) (Kohn 1999). Hence, a different way of thinking about rewards is required.

History has not helped; the twentieth century saw a dramatic shift in the type of work carried out by employees in the western world with a move away from large numbers of people carrying out repetitive manual labour-intensive tasks in facto-ries, yet assumptions amongst some senior management about motivation and engagement remain resolutely with the mass production ethos. Today's workforce is not dominated by large numbers of production line labour forces but increasingly by autonomous knowledge workers and creative-minded individuals who are looking for innovative solutions to undefined problems. In situations where creative thought is required, rewards not only do not work but the evidence points to the fact that extrinsic rewards (in particular financial rewards) can actually negatively affect performance (Ariely et al. 2009 cited in Pink 2009).

Incentives can introduce a competitive element which can fracture previously strong interpersonal relationships; teams of employees can splinter in the scramble for rewards, destroying cooperation (Kohn 1993). The negative effects of poorly thought-out and implemented incentive schemes on team dynamics are something the authors have observed with disputes and disruptions arising amongst previously harmonious teams.

Issues can also be hidden by employees if they believe revealing problems may have negative consequences, potentially leading to incentives being withheld (Kohn 1993). Incentives can also discourage risk-taking behaviour and potentially encourage people to engage in 'tweaking' of numbers and data to maintain previous incentive levels. A certain amount of 'massaging' of the numbers to achieve required targets is a temptation.

Incentives also have the potential to make managers lazy (Kohn 1993). It is far easier to dangle a reward in front of an employee than to spend time and effort in providing the support and encouragement for employees, ensuring they have a good job to do and are intrinsically motivated and rewarded by achievement and recog-nition. While in our experience the managers we have observed may not have been 'lazy', incentives certainly can be a distraction and can often (unintentionally) drive unsupportive management behaviour, causing them to focus on the incentive rather than the ethos (e.g. customer service, quality) behind it.

Evidence against poorly thought-out and implemented incentives is clear; how-ever, no claim is made that salary is unimportant. Our own experience, the literature and the mass media show that if wages are perceived to be unfair, it can have a substantial negative effect on people. Pay is of fundamental importance to employees. A fair market rate of pay, and if possible stretching beyond the market rate, will help attract and retain the best recruits. To put it simply, a fair rate will remove potential distractions and *take it off the table* (Pink 2009).

13.5.4 Measures and Targets (and a Quick Word About Goals)

It is vital that targets and measures are addressed as the impact on employees can be substantial. When considering employee engagement, a 'good' measure needs to take into account numerous aspects of both the employee's and the business' performance. Good measures provide a framework within which to operate so that everyone is clear on the part they need to play.

Call Handling Time (CHT) is a traditional measure widely used in contact centres. It is necessary to understand the average length of call, along with call volumes, for resource management purposes, but from a customer experience and employee engagement perspectives, it fails. A positive customer experience is more likely to be driven by successful resolution of an issue or an effortless transaction (Dixon et al. 2010), and employees are more likely to be motivated by successfully helping customers reach these goals [the 'emotional labour' component of customer service (Hochschild 1983; Millet et al. 2005)]. Furthermore, the random nature of calls means that the contact centre advisor has no control over how long any particular call can take and so being measured against CHT can cause stress and disengagement. Finally, the behaviours that may result from trying to meet CHT targets often impact other aspects of the process: vital information for field operations may be misreported or omitted altogether, and customers may be left uncertain of what will happen next.

A good measure is, therefore, one that meets several criteria:

- A holistic view of the process needs to be taken so that the business is able to meet its primary goal of serving the customer.
- The operative needs to understand the impact they can have on the measure and their value to the business.
- The operative needs to be able to take action to influence the measure.
- The measure needs to enable and support decision-making.

Measures are important for understanding business performance, but it is essential to understand that processes operate with a certain degree of randomness, causing variation beyond the employee's (and even business') control. By understanding that there is variation, rewarding or penalising employees for random events can be avoided (Deming 1982). Similarly, understanding exceptional levels of performance (both good and poor) can lead to better, more robust processes and improved business understanding.

Targets, as opposed to measures, often drive undesirable behaviour amongst management and employees and should be avoided. They also fail to look at the bigger picture. Goodhart's Law (1975), developed for use within social and economic policy, states that when a measure becomes a target, it ceases to be a good measure because targets and statistics can potentially hide the bigger picture and actually fail to measure what is intended, making the target itself invalid. For example, the ambulance service's 'speed to dispatch' target was identified by the National Audit Office (2011) as problematic because it resulted in multiple responses to incidents and unnecessary journeys. The 8-min target, designed for the most seriously ill patients, failed to take into account the broader but vital view of the well-being of all patients.

Becoming fixated on the target (and the underlying reward/punishment) rather than the ethos behind it is a common problem throughout many organisations. For example, focusing on meeting customer appointment slots can drive the inappropriate behaviour of failing to fulfil the appointment altogether. An organisation may meet their customer appointment target, but some customers may be left waiting for service if an employee acts to avoid the negative consequences of a missed appointment slot rather than be a few minutes late.

Lastly, goals have potential to improve employee performance. While goals can have a limited impact with highly complex and creative tasks, Locke and Latham (2002) found that setting employees' challenging (but not impossible) goals consistently leads to higher performance. Goals positively affect performance because they:

- Direct attention and effort within employees
- Have an energising function
- Encourage persistence (when employees have the necessary level of autonomy)
- Encourage interest and discovery, leading to the use of relevant knowledge

Motivation to achieve the goal amongst employees is raised if the goal has purpose or rationale, rather than a simple 'just do this' (see more in Job Design). Goal commitment is also enhanced by self-efficacy, which can be raised by training and experience. Finally, for goals to be successful, feedback is required so employees can understand progress towards the goals and alter their behaviour accordingly (Locke and Latham 2002).

13.5.5   Knowledge and Relevant Data

Any organisation needs to know how it is performing in a variety of areas, which can be summarised as financial, customer, process and learning and growth (Kaplan and Norton 1992). This information needs to take a holistic view of the business, with an end-to-end perspective of operations. But high-level, business-wide measures do not engage employees as they are often not relevant to an individual's job. Here, it is much more important to have a few key pieces of information that an individual can relate to: how it impacts on the customer and how they can use it and its value to the business.

Given the amount of information that any one person can process at a given time is limited (Sweller 1988), whether it is the number of systems that are being simultaneously operated or the number of tasks that a person needs to plan, it is necessary to restrict the data that an individual has to prioritise and manage so as not to cause stress through cognitive overload.

This is not to say that data such as task volumes, throughput and errors should not be collected; it is vital to understand how service and performance can be improved. It is essential to give employees a view of their performance. It is not always possible to provide instant feedback but timely updates, whilst particular events are still fresh in the memory and are important. Visible feedback is relatively easy to provide for static (or site-based) workers in the form of notice boards, control charts, etc. The challenge with the mobile workforce is to present timely, relevant information.

Similarly, an engaged employee is more likely to want to share information with their colleagues, and this should be encouraged and facilitated. It is easier to share information when employees are co-located, so the problems of knowledge sharing faced by mobile workforces need to be understood. Recent advances in technology allow smartphone apps and social networking sites to be used, enabling individuals to collect, share and receive information. Even so, the traditional 'water-cooler' conversation is often the most productive way of sharing knowledge, and mobile workers need to have opportunities to meet with colleagues.

13.5.6   Continuous Improvement

One way to engage a workforce is to give them a significant investment in improvement programmes. The scale of the programme can be either company wide or just within a department, but by inviting suggestions from the people that know the systems and processes best may not only yield the most enlightening solutions but also gain buy-in to the new solution.

Continuous improvement is an integral part of the Lean Manufacturing (Womack and Jones 2003) and Lean Service (Seddon 2005) philosophies, and the introduction of 'quality groups', equipped with root cause analysis skills, into workforces has been seen by these authors to have benefits ranging from teams adopting radically different roster patterns to meet demand levels to manyfold improvements in cycle times.

By providing the workforce with the autonomy to make decisions where they can see benefit both to the customer and the organisation, this can reap dividends for both the individual and the business. It is essential that there are no limits to the ideas put forward: a holistic and long-term view should be encouraged. The use of red-amber-green processes can assist with this freethinking. Each part of the process is coded accordingly: red if the process must be adhered to (e.g. for legal, regulatory or safety reasons), amber if the process can be changed but only after the benefits of improvement suggestions have been assessed and green if the process can be operated in the way that the employee seems to be most appropriate.

13.5.7   Job Design

'If you want people to do a good job, you give them a good job to do.' (Herzberg)

Giving employees meaningless, repetitive, dull tasks which have little variety can all but destroy engagement and morale, turning employees into robots who just 'do' rather than 'think'. In our experience, problems with engagement can also stem from employees who are required to just fix the symptom rather than solving the underlying root cause of the issue. Repetitive tasks, performed day in, day out, are unlikely to create a positive working experience or help engage employees. When employees are given a poorly designed job and then made to feel that their work is of little value, this is the ultimate kick the teeth to an individual's sense of worth and engagement.

Wanting to contribute to something which is meaningful is a basic human desire. Therefore, motivating and engaging employees is about enabling progress in work that has a purpose (Amabile and Kramer 2011; Pink 2009). Good job design needs employees to experience meaningful work and to enable this requires (Hackman and Oldham 1976):

- The opportunity and the ability to use a variety of skills
- The ability to observe and understand the outcome of their work efforts
- Participation in work which has an impact on others (including co-workers and customers)

It is about enriching the content of jobs with more opportunities for growth rather than just making the workload larger (Herzberg 1968).

Good job design also requires mechanisms for effective feedback. This includes providing employees with regular opportunities to garner information on what they have achieved and how they are contributing to the organisation, thus providing essential knowledge on their performance (Hackman and Oldham 1976). Feedback gives employees encouragement, direction and information and helps provide meaning and purpose to their work. In our experience when feedback is coupled with information on the value of the employee's contribution to the organisation, it provides a mechanism for positive reinforcement because it makes the employee feel valued by the organisation in return; it is a two-way street.

13.5.8 Career, Mastery and Advancement

Strongly related to job design is the need for employees to be able to advance and develop their career. The aspiration to be involved in something that matters and to get better at it, known as mastery (Pink 2009), is a key driver behind motivation and is aligned with the human need to advance, grow and attain fulfilment (Maslow 1943).

Inability or lack of opportunity to progress within organisations was found to be a demotivator (Herzberg 1968). In our experience a lack of career opportunities within workforces is linked to lower levels of morale and engagement. In some cases where opportunities were available, the opportunities were so infrequent they were effectively absent, colloquially known as *dead men's shoes*. This results in frustrated, disengaged employees who are more likely to tread water until other opportunities arise.

Hence, all organisations need to provide well-structured career paths and opportunities for employees to develop. Whilst it is true that some individuals may not seek promotion or development opportunities, most people need to know that the opportunity is there if required. It is also key for attracting the highest calibre recruits and retaining ambitious and talented employees.

13.5.9 Work Environment

What is considered by employees to be their 'place of work' can vary and certainly should not be restricted to the clean, bright and slightly sterile office that many associate with working environments. Some employees enjoy the isolation afforded by individual mobile working; others relish having a number of colleagues in close proximity. Open plan spaces can be distracting to some, but others may find the buzz stimulating and conducive to productive work. Domestic dwellings are also the working environment of a large number of mobile workers, as well as (more importantly) being the customer's home.

Amongst Herzberg's hygiene factors (1968) is the expectation that the working environment is of a satisfactory level. Being a hygiene factor, working conditions are more likely to upset and demotivate employees if they are poor or considered to be unacceptable, rather than have a strong positive effect on motivation if they are of a high standard. A poor work environment caused by, for example, heating and air conditioning problems not only causes complaints and potentially distracts employees but can also have the undesired effect of undermining the individual's sense of value to the organisation. Equally, mobile workers who have no fixed workplace and lack access to basics such as hot drinks or well-maintained toilets can feel unappreciated, even demoted, especially if such facilities have been available in the past.

Whereas a poor work environment can disengage employees, a good working environment can facilitate how 'valued' employees feel. The authors have spent time with mobile engineers who enjoy the autonomy of their role and not being under constant supervision; others enjoy the ability to have a 'picnic' style break (often with a view of fields) whilst they eat lunch. They often consider this to be part of their 'rights' or their 'psychological contract' with the company (Lester and Kickul 2001) and can afford a higher level of job satisfaction. Removing previously enjoyed 'rights' will potentially cause low morale and disenchantment.

Work environment is one area often overlooked in terms of employee engagement by organisations. Getting it right will help engender an employee's sense of worth to an organisation and consequently improve employee motivation and engagement.

13.6 Conclusions

'Be the change you wish to see.' (Gandhi)

Within this chapter the authors have described nine elements or components of employee engagement that they believe are fundamental to realising the full potential of a workforce. These components have been identified through the authors' fresh perspective on employee engagement using a blend of tried and tested research methods: the combination of qualitative methodologies, to reveal key insights into employees' attitudes and behaviours, with quantitative techniques, to provide essential data and information on the employee's job in hand. As a result a distinctive and revealing insight into employee engagement has emerged.

To close, a final word about culture and employee engagement and in particular about those driving the change. The prevailing culture within any organisation wishing to engage employees must be one of openness and respect for its people. Respect goes hand in hand with valuing employees, but also required is the ability to engender trust, often via honest and open discussions. Creation of trust is paramount; otherwise, concerns will not be aired, and progress towards change will be limited. The impetus to change also includes the values, attitudes and behaviours of the senior managers and leaders who typically dominate the culture of any company. Without their drive and commitment, any move towards a culture of employee engagement will only ever scrape the surface and, at the most, be superficial. Senior management must be fully committed to engage and motivate employees, including 'walking the walk'. Only then will the full benefits of employee engagement to both the business and individuals be fully realised.

References

Amabile TM, Kramer SJ (2011) The power of small wins. Harv Bus Rev 89(5):70–80

Ariely D, Gneezy U, Loewenstein G, Mazar N (2009) Large stakes and big mistakes. Rev Econ Stud 76(2):451–469

Brower MF (1995) Empowering teams: what, why, and how. Empower Organ 3(1):13–25

CBI report. Healthy returns? Absence and workplace health survey (2011) http://www.cbi.org.uk/media/955604/2011.05-healthy_returns_absence_and_workplace_health_survey_2011.pdf

Deci EL (1975) Intrinsic motivation. Plenum, New York, NY

Deci EL, Ryan RM (1985) Intrinsic motivation and self-determination in human behavior. Springer, Berlin

Deming WE (1982) Out of the crisis. Massachusetts Institute of Technology. Center for Advanced Engineering Study, Cambridge

Dixon M, Freeman K, Toman N (2010) Stop trying to delight your customers. Harv Bus Rev 88 (7/8):116–122

Goodhart CA (1975) Monetary relationships: a view from Threadneedle Street. In: Papers in monetary economics 1

Hackman JR, Oldham GR (1976) Motivation through the design of work: test of a theory. Organ Behav Hum Perform 16(2):250–279

Hardy B (2010) Morale: definitions, dimensions and measurement. PhD Thesis, University of Cambridge, Judge Business School

Harter JK, Schmidt FL, Hayes TL (2002) Business-unit-level relationship between employee satisfaction, employee engagement and business outcomes: a meta-analysis. J Appl Psychol 87(2):268–279

Haynes L, Service O, Goldacre B, Torgerson D (2012) Test, learn, adapt: developing public policy with randomised controlled trials. Cabinet Office Behavioural Insights Team, https://www.gov.uk/government/publications/test-learn-adapt-developing-public-policy-with-randomised-controlled-trials

Herzberg F, Mausner B, Snyderman BB (1959) The motivation to work. Wiley, New York

Herzberg F (1968) One more time: how do you motivate employees? Harv Bus Rev; 46–57

Hochschild AR (1983) The managed heart: commercialization of human feeling. University of California Press, Berkeley, CA

Kaplan RS, Norton DP (1992) The balanced scorecard–measures that drive performance. Harv Bus Rev 70(1):71–79

Kohn A (1993) Why incentive plans cannot work. Harv Bus Rev 71(5):54–63

Kohn A (1999) Punished by rewards: the trouble with gold stars, incentive plans. A's, praise, and other bribes. Mariner Books, New York

Lester SW, Kickul J (2001) Psychological contracts in the 21st century: what employees value most and how well organizations are responding to these expectations. Hum Resource Plann 24 (1):10–21

Locke EA, Latham GP (2002) Building a practically useful theory of goal setting and task motivation: A 35-year odyssey. Am Psychol 57(9):705

MacLeod D, Clarke N (2011) Engaging for success: enhancing performance through employee engagement, a report to Government

Maslow AH (1943) A theory of human motivation. Psychol Rev 50(4):370

McGregor D (1960) The human side of the enterprise. McGraw-Hill Professional, New York

Millet CT, Johnson SJ, Cooper CL, Donald IJ, Cartwright S, Taylor PJ (2005) Britain's most stressful occupations and the role of emotional labour. In: BPS occupational psychology conference, Warwick

National Audit Office (2011) Transforming NHS ambulance services. http://www.nao.org.uk/ambulance-service-2011.

Nohria N, Groysberg B, Lee LE (2008) Employee motivation. Harv Bus Rev July–Aug; 78–84

O'Reilly K (2004) Ethnographic methods. Routledge, New York

Oakland JS (1996) Statistical process control. Butterworth-Heinemann, UK

Pande PS, Neuman RP, Cavanagh RR (2000) The six sigma way. McGraw-Hill, New York

Pink D (2009) Drive—the surprising truth about what motivates us. Canongate Books, Edinburgh

Reichheld F (2006) The ultimate question: driving good profits and true growth. Harvard Business School Press, Boston, MA

Rucci AJ, Kirn SP, Quinn RT (1998) The employee-customer-profit chain at sears. Harv Bus Rev Jan–Feb; 82–97

Ryan RM, Deci EL (2000) Intrinsic and extrinsic motivations: classic definitions and new directions. Contemp Educ Psychol 25(1):54–67

Seddon J (2005) Freedom from command and control: a better way to make the work work; the Toyota System for Service Organizations. Vanguard Education Limited, New York

Shaw C, Ivens J (2002) Building great customer experiences. Palgrave Macmillan, New York

Strauss A, Corbin J (1994) Grounded theory methodology. Handbook of qualitative research. Sage, Thousand Oaks, CA, pp 273–285

Sweller J (1988) Cognitive load during problem solving: effects on learning. Cogn Sci 12 (2):257–285

Tamkin P, Cowling M, Hunt W (2008) People and the bottom line. Institute for Employment Studies (cited in D. MacLeod and N. Clark 2011)

Wilson PF, Dell LD, Anderson GF (1993) Root cause analysis. Quality, Milwaukee, WI

Womack JP, Jones DT (2003) Lean thinking: banish waste and create wealth in your corporation. Free Press, New York

Chapter 14
The Asset Replacement Problem State of the Art

Amir H. Ansaripoor, Fernando S. Oliveira, and Anne Liret

Abstract This book chapter outlines the different modelling approaches for realising sustainable operations of asset replacement and studying the impact of the economic life, the repair-cost limit and comprehensive cost minimisation models. In particular it analyses in detail the parallel replacement models and suggests a new model that addresses some of the issues not yet solved in this area. Finally a discussion about the limitations of the current models from a theoretical and applied perspective is proposed and identifies some of the challenges still faced by academics and practitioners working on this topic.

14.1 Introduction

As assets age, they generally deteriorate, resulting in rising operating and mainte-nance (O&M) costs and decreasing salvage values. Moreover, newer assets that have a better performance and keep better their value may exist in the marketplace and be available for replacement. Therefore, public and private organisations that maintain fleets of vehicles, and/or specialised equipment, need to decide when to replace vehicles composing their fleet. These equipment replacement decisions are usually based on a desire to minimise fleet costs and are often motivated by the state of deterioration of the asset and by technological advances (Hartman 2005).

The general topic of equipment replacement models was first introduced in the 1950s (Bellman 1955). By using dynamic programming, Bellman developed a

A.H. Ansaripoor
ESSEC Business School, Singapore, Singapore

F.S. Oliveira
ESSEC Business School, Paris, France

A. Liret (✉)
Research and Innovation, BT Technology, Services and Operations, Paris, France
e-mail: anne.liret@bt.com

G. Owusu et al. (eds.), *Transforming Field and Service Operations*,
DOI 10.1007/978-3-642-44970-3_14, © Springer-Verlag Berlin Heidelberg 2013

model in order to obtain the optimal age of replacement of the old machine with a new machine. Another important subject was the development of parallel replacement models in which management decisions are made for a group of assets instead of one asset at the time (Hartman and Lohmann 1997).

Vehicle replacement is a key role of fleet provisioning teams. Indeed field service operational planning and delivery primarily relies on the assumption that the whole engineering force can be furnished with the vehicle appropriate for the service, at any time. In practice, the choice of the adequate type, brand, and technology depends on internal factors (such as the engineer role and service environment, but not systematically mileage driven) and on external factors (such as fuel price variation, government carbon emission incentives, manufacturing costs, and maintenance costs). Moreover, in addition to risk and field force efficiency, the impact of vehicle replacement on customer experience needs to be considered as well. This suggests a twofold fleet planning problem that vehicle replacement aims to address: a planned fleet portfolio and a rental plan for jeopardy situations.

In addition, field service enterprises face increasing challenges on carbon emissions and cost reduction. This need to transform the way field services operate has an impact on the choice of vehicles within a business, affecting the vehicle replacement processes. When attempting to optimise the fleet composition, which is essential for achieving sustainability, we need to take into account several factors (some of which are stochastic and uncertain in nature), which need to be addressed before low-carbon vehicles are a feasible alternative for field service operations including the intangible reputation of sustainable energy investment, the evolution of market prices, strategic partnerships, and risk sharing.

This chapter aims at outlining the historical developments of the asset replacement problem, discussing the limitations of the models developed so far, and introducing a new model which overcomes some of these drawbacks. Section 14.2 presents a classification of different asset replacement models, which are broadly categorised into serial and parallel models. Section 14.3 describes the different approaches to modelling the asset replacement problem. Section 14.4 discusses the methods used to solve the parallel asset replacement problem and suggests a new formulation to address some of their drawbacks. Section 14.5 outlines our analysis from literature review, with Sect. 14.6 emphasizing the challenges from a practical service industry perspective. Section 14.7 concludes this chapter.

14.2 The General Classifications of Fleet (Asset) Replacement Models

The models generally can be categorised into two main groups based on different fleet (asset) characteristics: homogenous and heterogeneous models. In the homogeneous replacement models, a group of similar vehicles in terms of type and age, which form a cluster, have to be replaced simultaneously (each cluster or group cannot be decomposed into smaller clusters). On the other hand, in the heteroge-neous model, multiple heterogeneous assets, such as fleets with different types of vehicle, have to be optimised simultaneously. For instance, vehicles of the same type and with the same age may be replaced in different periods (years) because of the restricted budget for procurement of new vehicles. The heterogeneous models are closer to the real-world commercial fleet replacing problem. These models are solved by integer programming, and, generally, the input variables are assumed to be deterministic (Hartman 1999, 2000, 2004; Simms et al. 1984; Karabakal et al. 1994).

The most popular methodology for solving homogenous models is dynamic programming. The advantage of the homogenous model is to take into account probabilistic distributions for input variables (Hartman 2001; Hartman and Murphy 2006; Oakford et al. 1984; Bean et al. 1984; Bellman 1955).

Another important classification of these models regards the nature of the replacement process: parallel vs. serial, e.g. Hartman and Lohmann (1997). The main difference between parallel replacement analysis and serial replacement analysis is that the former takes into account how any policy exercised over one particular asset affects the rest of the assets of the same fleet. An example of parallel replacement would be a fleet of trucks that service a distribution centre. In this case, the total available capacity is the sum of the individual capacities of the trucks. In the serial replacement model, the assets operate in series, and consequently, demand is satisfied by the group of assets which operate in sequence. An example of this case is a production line in which multiple machines must work together to meet a demand or service constraint. In general, the capacity of the system is defined by the smallest capacity in the production line (Hartman 2004).

The following definition of parallel replacement comes from (Hartman and Lohmann 1997). Parallel replacement deals with the replacement of a multitude of economically interdependent assets which operate in parallel. The reasons for this economic interdependence are:

1. Demand is generally a function of the assets as a group, such as when a fleet of assets are needed to meet a customer's demands.
2. Economies of scale may exist due to purchasing assets and promoting large quantity of purchases.
3. Diseconomies of scale may exist with maintenance costs because assets which are purchased together tend to fail at the same time.
4. Budgeting constraints may require that assets compete for available funds. These characteristics, either alone or together, can cause the assets to be economically interdependent.

On the other hand, the serial replacement analysis assumes a certain utilisation level for an asset throughout its life cycle. Hartman (1999) mentioned that since utilisation levels affect operating and maintenance costs and salvage values (which in turn influence replacement schedules), a replacement solution is not optimal unless utilisation levels are also maximised. This suggests a strong dependency relationship between asset utilisation levels and the combination of demand requirements, the number of assets available, and the capacity of each asset.

Next section presents different approaches to modelling the asset replacement problem: the economic life cycle, the repair-cost limit, the comprehensive cost minimisation, and the issue of decreasing utilisation with age.

14.3 Approaches for Replacement Decisions

The goal driving a replacement decision consists of identifying replacement candidates among fleet or asset members so that the total costs are minimised in the long run. In this section we review different approaches for deciding the optimal time for candidate asset replacement.

14.3.1 Approaches Based on the "Economic Life"

An intuitive method for identifying replacement candidates is to use a replacement standard, such as the age of the equipment. For example, assets older than a standard threshold should be replaced. Additionally, a ranking profile can be used in order to sort the equipment units by how much they exceed the threshold. For example, Eilon et al. (1966) considered a model for the optimum replacement of forklift trucks. The parameters in their model were the purchase price, the resale value, and the maintenance costs of the equipment. The goal of their model was to derive the minimum average costs per equipment year, and the corresponding optimal equipment age policy, for a fleet of forklift trucks.

Let us now describe the model proposed by Eilon et al. (1966) in more detail. Let $TC(t)$ be the total average annual (or per period) cost of an existing truck, assuming it is replaced at age (time) t. Let A stand for the acquisition cost of new truck, $S(t)$ be the resale value of the existing truck at age t, $C(t)$ be the accumulated depreciation costs up to time t, τ be the rate of taxation, and $f(t)$ be the maintenance costs of a truck, t years after acquisition. Then the total average annual cost of an existing truck is represented by (14.1):

$$TC(t) = \frac{1}{t}(A - S(t) - C(t).\tau) + \frac{1}{t}\int_0^t f(t)dt \qquad (14.1)$$

The first term in (14.1) represents the average capital costs involved in the acquisition of the existing truck, taking into account the savings from resale value and tax savings from depreciation. The second term in (14.1) expresses the total average maintenance costs for the existing truck over the years up to the present time t. The minimum total average annual costs, as a function of t, determines the optimal replacement time.

The economic life of an asset (also known as service life or lifetime of the asset) is defined as the age which minimises the *equivalent annual cost* (EAC) of owning and operating the asset. The EAC includes purchase and *Operating and Maintenance* (O&M) costs minus salvage values. Generally, O&M costs increase with age while salvage values decrease with age. As a result, the optimal solution represents a trade-off between the high costs of replacement (purchase minus salvage) and increasing O&M costs over time.

The concept of economic life is easier to describe graphically. In Fig. 14.1, adapted from Hartman and Murphy (2006), it is assumed that the initial purchase cost is $100,000, with the salvage value declining 20 % per year. O&M costs are expected to increase 15 % per year after $11,500 in the first year. Figure 14.1 illustrates the annualised O&M and capital costs and their sum (EAC) for each possible of age assuming an annual interest rate 8 %. Once the optimal economic life is determined, the asset should be continuously replaced at this age, if we assume repeatability and stationary costs.

In order to obtain the EAC, when retaining an asset for n periods, all costs over the n periods must be converted into n equal and economically equivalent cash flows. Then, the economic life of an asset is typically computed by calculating the EAC of retaining an asset for each of its possible service lives, ages one through n, and the minimum is chosen from this set (Hartman 2005; Weissmann et al. 2003; Hartman and Murphy 2006).

Yatsenko and Hritonenko (2011) have also considered the economic life (EL) method of asset replacement taking into account the effects technological improvements which decrease maintenance costs, new asset cost, and salvage value. They have shown that, in general, the EL method renders an optimal replacement policy when the relative rate of technological change is less than one percent. However, for larger rates, they recommend annual cost minimisation over the two future replacement cycles, which was earlier proposed and implemented by Christer and Scarf (1994).

14.3.2 Approaches That Consider a Repair-Cost Limit

Another replacement criterion is the repair cost. When a unit requires repair, it is first inspected and the repair cost is estimated. If the estimated cost exceeds a threshold, which is known as "repair limit" then the unit is not repaired but, instead, is replaced. Repair limits have long been used and their values have often been based on the principle that no more should be spent on an item than it is worth. This criterion is indeed an important one. There is evidence that repair-cost limit policies have some advantages in comparison with economic age limit policies. For example, Drinkwater and Hastings (1967) analysed data for army vehicles. They obtained the repair limiting value in which the expected future cost per vehicle-

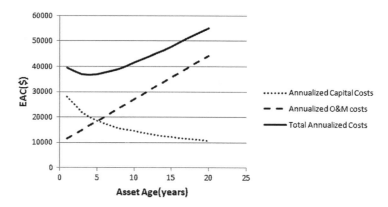

Fig. 14.1 Annualised purchase cost, O&M cost, and total (EAC) costs

year when the failed vehicle is repaired is equal to the cost in which the failed vehicle is scrapped and a new one is substituted. Specifically, they defined two options:

1. Repair the vehicle.
2. Scrap the vehicle and replace it by a new one. This is called a repair decision.

We now present the model used for the repair decision in more detail. We consider a vehicle at age t which requires repair. If we select option 1, to repair the vehicle, the future cost per vehicle-year is represented by (14.2) in which r is the present cost of repair, $c(t)$ is the expected total cost of future repairs, and $l(t)$ is the expected remaining life of the vehicle:

$$\frac{r + c(t)}{l(t)} \tag{14.2}$$

If we select option 2, scrapping the vehicle will incur an expected future cost per vehicle-year being δ, which is defined by the average cost per vehicle-year up to age t. Obviously, the repairing decision (option 1) will be selected if (14.3) holds; otherwise, the scrapping decision will be chosen. Therefore, the critical value of r is determined by (14.4) in which the future cost per vehicle-year equals the average cost per vehicle-year up to age t. As a result, the optimal repair limit at time t, $r^*(t)$, is determined by (14.5):

$$\frac{r + c(t)}{l(t)} < \delta \tag{14.3}$$

$$\frac{r^*(t) + c(t)}{l(t)} = \delta \tag{14.4}$$

$$r^*(t) = \delta l(t) - c(t) \tag{14.5}$$

Drinkwater and Hastings (1967) have shown that the repair-cost limit policy is better than the economic age policy. Nonetheless, there is a main drawback to the conventional repair-cost limit policy: the repair/replace decision is based only on the cost of one single repair. Under this condition, a system with frequent failures and, consequently, high accumulated repair costs will continue to be repaired rather than replaced. As a result, an improved policy making the repair/replace decision based on the entire repair history would be a better criterion. In order to address this issue, Chang et al. (2010) have developed a generalised model for determining the optimal replacement policy based on multiple factors such as the number of minimal repairs before replacement and the cumulative repair-cost limit. The main characteristic of their model is to consider the entire repair-cost history. Nakagawa and Osaki (1974) have also suggested an alternative approach which does not focus on repair costs but, instead, on repair time. If the repair process is not completed up to the fixed repair time limit, then the unit under repair is replaced by a new one. The repair time limit is obtained by minimising expected costs per unit of time over an infinite time horizon.

14.3.3   Comprehensive Cost Minimisation Models

There are other approaches that generalise the problem of optimal replacement by taking into account the optimal decisions for acquisition, operation, and replacement policies. For example, Simms et al. (1984) have analysed a transit bus fleet in which the equipment units in the fleet system were assigned to perform different tasks, at different levels, subject to changing capacity constraints. Their objective was to minimise the total discounted cost over a finite horizon.

14.3.3.1   Objective Function

The objective function is represented by (14.6), in which t and a are the indices for time periods (year) and age of the buses, respectively, and T is the length of the planning horizon, in years. The decision variables are the number of route kilometres travelled by a bus with age a, in year t, m_{ta}; the number of buses with age a, which operate in year t, x_{ta}; and the number of new buses which should be purchased, with an acquisition cost L_t, at the beginning of year t, denoted by p_t. In each year the price of selling a bus with age a is represented by S_{ta}, and $C_{ta}(m_{ta})$ is the cost of operating a bus with age a, in year t, for the associated kilometres travelled by m_{ta}. Finally, γ represents the discount factor. In (14.6) the first term represents the acquisition costs, the second term stands for the revenue received from selling the buses, and the third term denotes the cost of operating the buses.

Simms et al. (1984) computed the optimal acquisition, operation, and selling policies using dynamic programming:

$$\underset{m_{ta}, x_{ta}, p_t}{Min\ Z} = \sum_{t=0}^{T} \gamma^t p_t L_t - \sum_{t=0}^{T} \gamma^{t+1} \sum_a \left(x_{ta} - x_{t+1,a+1} \right) S_{t+1,a+1}$$

$$+ \sum_{t=0}^{T} \sum_a \gamma^i x_{ta} C_{ta}(m_{ta}) \tag{14.6}$$

On the same topic, Hartman (1999) has considered the replacement plan and corresponding utilisation levels for a multi-asset case in order to minimise the total cost. He generalised equipment replacement analysis as it explicitly considers utilisation as a decision variable. His model allows assets to be categorised according to age and cumulative utilisation while allowing their periodic utilisation to be determined through analysis. As a result, he has considered simultaneously tactical replacement and operational decisions, taking into account the trade-offs between capital expenses (replacement costs) and operating expenses (utilisation costs). The objective was to minimise the total cost of assets that operate in parallel. He solved the problem using linear programming. Furthermore, Hartman (2004) has generalised this same problem by incorporating a stochastic demand. He solved the problem using dynamic programming. Overall, none of these approaches introduced any special new replacement criteria and only presented optimisation methodologies in order to minimise the cost of corresponding fleets.

14.3.3.2 Modelling Fleet Life Conditions

Following the model proposed by Simms et al. (1984), the nonlinear constraint (14.7) requires the fleet to drive a minimum value of total kilometres per year, M_t. Constraint (14.8) expresses the boundary conditions for the decision variable m_{ta}, in which m_- and m_+ respectively denote the minimum and maximum number of kilometres that a single bus can drive in a given year. Constraint (14.9) represents the requirement that at least a minimum number of buses, N_t, in each year, should be in the fleet:

$$\sum_a x_{ta} m_{ta} \geq M_t \quad \forall t \in \{0, 1, 2, \ldots, T\} \tag{14.7}$$

$$m_- \leq m_{ta} \leq m_+ \tag{14.8}$$

$$\sum_a x_{ta} \geq N_t \quad \forall t \in \{0, 1, 2, \ldots, T\} \tag{14.9}$$

In inequality (10), Q is the minimum age for a bus to be considered for a sell decision and the left-hand side is equal to the number of buses which are sold at the beginning of the corresponding year. Therefore, inequality (10) stands for a consistency constraint, in the sense that it does not permit old buses to be bought:

$$x_{ta} - x_{t+1,a+1} \geq 0 \quad , a \geq Q - 1 \tag{14.10}$$

Equation 14.11 means that the buses are not eligible for sale until their reach to the minimum age Q. Equation 14.12 represents the boundary conditions, in which K_a are the initial numbers of buses for the different ages:

$$x_{ta} - x_{t+1,a+1} = 0 \quad , a < Q - 1 \tag{14.11}$$

$$x_{(-1)a} = K_a \quad , x_{(T+1)j} = 0 \tag{14.12}$$

If budget constraints for capital acquisitions are also considered, then the constraint (14.13) are also required, in which B_t is the capital budget in period t. Furthermore, if there is also an operating budget constraint, then we also need to impose constraint (14.14) in which O_t is the operating budget in period t:

$$p_t \leq \frac{B_t}{L_t} \tag{14.13}$$

$$\sum_a x_{ta} C_{ta}(m_{ta}) \leq O_t \tag{14.14}$$

The model represented by (14.6–14.14) has a nonlinear objective function subject to a set of nonlinear constraints. By using dynamic programming, Simms et al. (1984) solved the problem. If we compare the two models proposed by Simms et al. (1984) and Keles and Hartman (2004), we understand that regardless of the solving methodology used, the main difference is considering the behaviour of utilisation as a function of age of the vehicles and assuming it as a decision variable by Simms et al. (1984). Another difference is that Simms et al. (1984) considered the same type of asset, whereas Keles and Hartman (2004) considered multiple types of asset. However, for the rest of the components of the two models, i.e. the goal of the objective function and the constraints, they are almost the same.

Another important issue that requires particular modelling attention is the relation between age and utilisation. The utilisation intensity (annual mileage) of vehicles exploited by transportation companies decreases with time of exploitation/ cumulative mileage probably in real-life cases. The youngest vehicles are usually utilised more intensively than the oldest ones, because their unit exploitation costs are lower (e.g. fuel consumption is lower), and the depreciation costs could be ignored. Examples of the occurrence of such pattern can be found in Kim et al. (2004) and Simms et al. (1984), and it fits well with real-world situations. This pattern can cause an issue, in particular in bus fleet management (Simms et al. 1984), because if the relation associating utilisation with age is not considered, one would expect that the older buses would be replaced first and younger buses kept. However, in practice, this is not the case and does not appear systematically

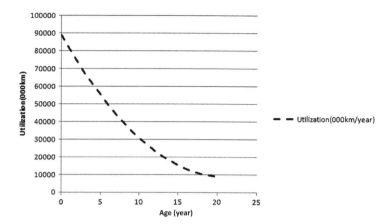

Fig. 14.2 Annual utilisation by age

suitable for two reasons. Firstly, older buses are usually kept only to meet peak daily demand and these buses accumulate only the minimum number of route kilometres during the year. Secondly, the resale value of younger buses is much higher than older buses. Therefore, even if the operating cost of older buses is higher, this is compensated by the fact that they do operate minimal route kilometres: the extra expense is lower than the gain obtained by selling younger buses. So, this suggests as a meaningful assumption to distinguish two levels of utilisation for an urban transit bus fleet with different ages. Simms et al. (1984) concluded that a high utilisation level is considered for buses with less than 10 years for satisfying the normal demand and a low utilisation level for buses more than 10 years in the case of peak demand. Figure 14.2 illustrates this vehicle utilisation pattern in a real-world field service situation.

Redmer (2009) has also considered the relationship between utilisation intensity and ageing by applying the minimal average cost replacement policy using the following considerations:

- The utilisation intensity (annual mileage) of vehicles for each year of their operational life has to be taken into account.
- The vehicles' exploitation costs have to be divided into fixed costs (independent of utilisation intensity but varying with time of exploitation/cumulative mileage), running costs (depending on utilisation intensity/mileage and varying with time of exploitation/cumulative mileage), and fuel costs (varying with time of exploitation/cumulative mileage).
- The total costs of exploitation and ownership have to be given per 1 km or mile.
- The technical durability of vehicles (e.g. maximal mileage) has to be taken into account.
- Different forms of financing the fleet investments (buying for cash, credit, leasing, and hiring) have to be considered.

Redmer (2009) also outlined the advantages and issues of solving different replacement strategies applied in parallel. Next section focuses on modelling the parallel replacement problem and proposes a new model for addressing the issues.

14.4 The General Parallel Replacement Problem

In this section we commonly refer to groups of assets as fleets. However, the model is general in the sense that cost functions are specified without operational details. Thus, this analysis may be applied to a manufacturing setting if the costs can be quantified. The parallel replacement models are usually difficult to solve due to their combinatorial nature as mentioned by Hartman (2000), leading to hypothesis making on the general statement. Jones et al. (1991) considered a parallel replacement problem on the condition of fixed replacement costs. Rajagopalan (1998) and Chand et al. (2000) have proposed dynamic programming algorithms that simultaneously consider the replacement and capacity expansion problems.

14.4.1 An Integer Programming Formulation of the Parallel Replacement Problem

Given the complex nature of the problem, the case of multiple alternatives within parallel replacement has been rarely considered in the literature. However, using an integer programming formulation, it is possible to deal with multiple choices under economies of scale and budgeting constraints (Keles and Hartman 2004).

- **Objective function**: The objective function represents the costs associated with each challenger's discounted cash flows which are purchasing, operating, and maintenance costs subtracting the revenue from salvage values. The objective function is summarised in (14.15). All costs in the model are assumed to be discounted to time zero using an appropriate discount rate. The fixed cost associated with asset buying is represented by f_t. The new asset acquisition cost per unit in each year is l_{it}. The operating and maintenance cost is shown by c_{iat}, and the salvage revenue is represented by r_{iat}. I represents the total number of challengers (i.e. available alternatives for assets) in each period. The maximum age of any asset associated with its type is shown by A_i, and the length of time horizon is assumed to be T—typically T is assumed to be less than 15 years:

$$\underset{X,S,Z}{Min} \sum_{i=1}^{I} \left[\sum_{t=0}^{T-1} \left(f_t Z_t + \sum_{a=0}^{A_i-1} l_{it} X_{i0t} \right) + \sum_{t=0}^{T-1} \sum_{a=0}^{A_i-1} c_{iat} X_{iat} - \sum_{t=0}^{T-1} \sum_{a=1}^{A_i} r_{iat} S_{iat} \right] \quad (14.15)$$

- **Decision variables**: The total number of assets which are currently used in the system is represented by X_{iat} $(a > 0)$. The variable indices are a, t, and i which stand for the age of the assets (buses), time periods, and type of the assets, respectively. The decision variables are the number of the assets bought at the beginning of each year, X_{i0t}, the number of assets which are salvaged at the end of each year, S_{iat}, and a binary variable confirming an acquisition in year t, Z_t.
- **Constraints**: Constraint (14.16) states that enough assets (or capacity) have to be available to satisfy demand for buses at time t, d_t:

$$\sum_{i=1}^{I} \sum_{a=0}^{A_i-1} X_{iat} \geq d_t \quad \forall t \in \{0, 1, \ldots, T-1\} \quad (14.16)$$

- Equation 14.17 represents the capital budgeting constraint to limit the payment for new asset acquisitions with predetermined capital budget, b_t, in each year:

$$\sum_{i=1}^{I} \sum_{a=0}^{A_k-1} l_{it} X_{i0t} + f_t Z_t \leq b_t \quad \forall t \in \{0, 1, \ldots, T-1\} \quad (14.17)$$

- Constraint (14.18) describes that the initial number of assets, h_{ia} $(a > 0)$, should be either used, X_{ia0}, or salvaged, S_{ia0}. Equation 14.19 shows that the number of used assets in 1 year should be either used or salvaged in the next year:

$$X_{ia0} + S_{ia0} = h_{ia} \quad \forall \; a \in \{1, 2, \ldots, A_k\}, \forall i \in I \quad (14.18)$$

$$X_{i(a-1)(t-1)} = X_{iat} + S_{iat} \quad \forall i \in I, \; \forall \; a \in A_i, \forall t \in \{1, 2, \ldots, T\} \quad (14.19)$$

- Constraint (14.20) requires that all assets should be sold in the last year of the planning horizon (T). Equation 14.21 presents that any asset that has reached its maximal age is not used anymore:

$$X_{iaT} = 0 \quad \forall\, a \in \{0, 1, 2, \ldots, A_i - 1\} \tag{14.20}$$

$$X_{iA_i t} = 0 \quad \forall\, i \in I,\ \forall\, t \in \{0, 1, 2, \ldots, T\} \tag{14.21}$$

- Constraint (14.22) prohibits salvaging any new asset immediately. Indeed, for salvaging of any new purchased asset at least one year should be passed. Finally, constraint (14.23) requires non-negative, integer solutions:

$$S_{i0t} = 0 \quad \forall\, i \in I,\ \forall\, t \in \{0, 1, 2, \ldots, T\} \tag{14.22}$$

$$X_{iat}, S_{iat} \in \{0, 1, 2, \ldots\},\ Z_j \in \{0, 1\} \tag{14.23}$$

Solving the model represented in (14.15–14.23) provides quantitative data. An extensive sensitivity analysis, fed with this data, is generally required when we want to consider the impact of various parameters on the optimal policies and finally choose the appropriate type and timing for bus replacement.

The aforementioned papers on the parallel replacement problem were considered in a deterministic framework. Replacement models in the case of existence of uncertainty were focused mainly on single or serial replacement problems. For example, Ye (1990) presented a single replacement model in which operating costs and the rate of deterioration of equipment were stochastic and the optimal time for replacing was determined in a continuous-time setting. Dobbs (2004) developed a serial replacement model in which operating costs were modelled as a geometric Brownian motion and the optimal investment time was obtained. Rajagopalan et al. (1998) developed a dynamic programming algorithm for the case where a sequence of technological breakthroughs was anticipated but their magnitude and timing were uncertain. A firm, operating in such an environment, should decide how much capacity of the current technology to acquire to meet future demand growth.

Parallel replacement model has been very successful in other types of applications. Feng and Figliozzi (2013) have considered a fleet replacement framework for comparing the competitiveness of electrical with conventional diesel trucks. They adapted the model described above to scenarios with different fleet utilisation and fuel efficiency. By using sensitivity analysis of ten additional factors, they have shown that electrical vehicles are more cost effective when conventional diesel vehicles' fuel efficiency is low and daily utilisation is above some threshold. Breakeven values of some key economic and technological factors that separate the competitiveness between electrical vehicles and conventional diesel vehicles were calculated in all scenarios.

Typically, in the comparison of the performance of electrical and conventional vehicles, one takes into account the high capital costs associated with electrical engine vehicles. The replacement decision depends on the result of a complete economic and logistics evaluation of the competitiveness of the new vehicle types.

In addition, as vehicles age, their per-mile operating and maintenance costs increase and their salvage values decrease. So, when the O&M costs reach a relatively high level, it may become cost effective to replace fossil fuel vehicles since the savings from O&M costs may compensate the high capital cost of purchasing new engine vehicles. Moreover, if fleet managers are enthusiastic in replacing conventional vehicles with new electric vehicles, it is important to understand how the O&M costs and salvage values change over time. Conventional diesel and electric commercial vehicles have significantly different capital and O&M costs.

14.4.2 A General Parallel Heterogeneous Asset Leasing Replacement Model

In this subsection we introduce a general asset replacement model for obtaining optimal replacement decisions regarding K types of assets under leasing framework. Specifically, a heterogeneous model is developed in which the assets are bounded by common budget constraints, demand constraints, and a fixed cost that is charged in any period in which there exist a replacement. It is assumed that in any period, assets from any of K types can be leased in order to replace retired assets for meeting corresponding demand in that period. The section ends with a customised variant for vehicles fleet.

The notation and formulation to be presented is more easily described by the network in Fig. 14.3. For the sake of simplicity, this figure represents the case of two asset types that are available to meet the demand ($I = 2$). The age of the asset in years, a, is defined on the vertical axis (maximum A), and the end of the planning period in years, t, is defined on the horizontal axis (horizon T). Due to the fact that we are considering a commercial setting, the leasing period is assumed to be 4 years. So, based on this assumption, the model is represented with $A = 3$ and $T = 6$. Indeed, at the end of time horizon $T = 6$, all the assets are retired.

Each node is defined according to the pair (a,t). The flow between these nodes, noted X_{iat}, represents an asset of age a in use from the end of time period t to the end of period $t + 1$, in which the asset is of age $a + 1$. Assets are either provided from the initial fleet, represented as flow from supply nodes n_{ia}, or must be leased, represented as X_{i0t} flow in each period t. An asset when reaches age A must be retired. All assets are retired at the end of the horizon. For meeting the associated demand in each period, the retired assets should be replaced by leasing new assets. In Fig. 14.3, the two types of assets are represented by different arcs (dashed or solid).

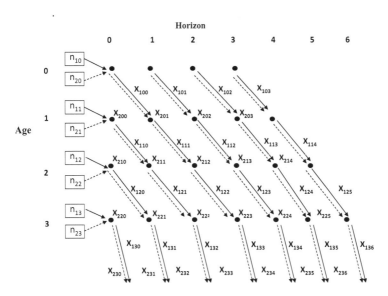

Fig. 14.3 Challengers are denoted by different arcs and different source (initial fleet) nodes. Nodes are labelled (a, t) with a being the age of the asset and t the time period. Flow X_{iat} represents asset leased $(a = 0)$ and assets in use $(a > 0)$

14.4.2.1   Exploiting Asset Portfolio for Fleet Replacement

Let us adapt the introduced model for fleet replacement. We consider two types of technologies: the fossil fuel technology (defender) and the new engine technology (challenger). Moreover, we take into account the leasing option for financing the commercial fleet investments, which is seen as the best option in the commercial setting by Redmer (2009). This leads to a deterministic model. Future economic and technical factors and costs, such as lease prices, fuel prices, fuel, and electricity consumption rates, are assumed to be known functions of time and vehicle type.

The indices in the model are the types of vehicle, $i \in \{1, 2\}$, the maximum age of vehicles in years, $a \in A$; $A = \{1, 2, \ldots, A\}$, and the time periods (year), $t \in T$; $T = \{0, 1, \ldots, T\}$. The decision variables include the number of type i, age a vehicles which are currently leased in year t, X_{iat}, and the number of type i vehicles which are leased at the beginning of year t, P_{it}. The parameters are:

- **The expected utilisation** (miles travelled per year) of a type i, age a vehicle in year t (miles/year), u_{iat}
- **The expected demand** (miles need to be travelled by all vehicles) in year t (miles), d_t
- **The available budget** (money available for leasing new vehicles) in the beginning of year t, b_t
- **The initial number of vehicles** of type i, age a at the beginning of first year, h_{ia}
- **The lease cost** of a type i vehicle, l_i

- **The expected operating (running) cost** – per mile – of a type i, age a vehicle in year t, o_{iat}
- **The emissions cost** – per mile – of a type i, age a vehicle, e_{ia}

The objective function which we want to minimise (14.24) is the sum of leasing costs for the period (T-3) and the operating (running) cost for the entire horizon to the end of year T:

$$Min \sum_{i=0}^{I} \sum_{t=0}^{T-3} (l_i P_{it}) + \sum_{i=0}^{I} \sum_{a=0}^{A} \sum_{t=0}^{T} [o_{iat} + e_{ia}] u_{iat} X_{iat} \qquad (14.24)$$

Equation 14.25 shows that the leasing costs cannot exceed the annual budget. Equation 14.26 requires that the total miles travelled by all used vehicles meet the annual demand:

$$\sum_{i=i}^{I} l_i . P_{it} \leq b_t \qquad \forall t \in \{0, 1, 2, \ldots, T-3\} \qquad (14.25)$$

$$\sum_{a=0}^{A} \sum_{i=i}^{I} X_{iat} u_{iat} \geq d_t \quad \forall t \in \{0, 1, 2, .., T-3\} \qquad (14.26)$$

Equation 14.27 describes that the total number of the vehicles with different ages and types in the first year should be equal to the initial condition of the system:

$$X_{ia0} = h_{ia} \quad \forall i \in I, \forall a \in A \qquad (14.27)$$

In addition, (14.28) shows that in the last 4 years of the planning horizon, there is no leasing of new cars. In (14.29) the number of new leased cars at the beginning of each year is determined:

$$P_{it} = 0 \qquad \forall i \in I, \forall t \in \{T-3, \ldots, T\} \qquad (14.28)$$
$$P_{it} = X_{i0t} \quad \forall i \in I, \forall t \in \{0, 1, 2, \ldots, T-3\} \qquad (14.29)$$

Equation 14.30 represents the flow equation in which the number of the cars at each year equals to the number of new leased cars plus the number of cars belonged to the previous year. Finally, expression (14.31) is the constraint for non-negative numbers of decision variables:

$$X_{iat} = P_{it} + X_{i(a-1)(t-1)} \quad \forall i \in I, \forall a \in A, \forall t \in T \qquad (14.30)$$
$$X_{iat}, P_{it} \in Z^+ \qquad (14.31)$$

Having analysed extensively the different models in the literature and identified some of their limitations, next, in Sect. 14.5, we summarise the main insights from our review of these different approaches.

14.5 Insights from the Literature on Fleet Replacement Models

The aforementioned replacement policies and methods represent only a small part of all efforts that have been done to solve the equipment replacement problem in general (Nakagawa 1984; Ritchken and Wilson 1990) and the vehicle replacement problem in particular (Eilon et al. 1966). The vehicle replacement policy has a prominent role in transportation companies and belongs to an important class of the fleet strategic management problems that have been extensively considered in the literature during last 50 years (Dejax and Crainic 1987). Nevertheless, there are many obstacles for applying the existing methods. Such obstacles exist from the following features of the existing replacement methods:

- Most of the methods are assumed to be applied in a stable environment which is not the case for most of the vehicles in under operational conditions, for example, the way those vehicles are utilised and the loads carried, the climate, and other factors from road conditions which can have impact on fuel economy of the vehicles.
- Focused on a given group (type) of vehicles, they do not go to the granularity of single vehicle.
- Assumptions taken such as a constant utilisation rate of the equipment during its operational life may be too far from field service real-world situations.

In practice, the existing models have at least one of the mentioned drawbacks. For instance, Eilon et al. (1966) consider particular vehicles but assume a fixed utilisation pattern, whereas Simms et al. (1984) relax the assumption of the constant utilisation but constrain an age to the replacement problem by placing a lower bound of 15 years. Suzuki and Pautsch (2005) constrain an age to the replacement model by putting an upper bound of 5 years and conclude that vehicles of age 6 or beyond may not be suitable for business operations: that contradicts the assumption of Simms et al. (1984). Moreover, the significant part of the vehicle replacement models assumes budget constraints (Simms et al. 1984). This part is actually important when replacement policy is defined for fleet of vehicles but not for particular vehicles. However, such constraints generally result in replacement of the limited group of the oldest vehicles (Redmer 2009). Because of the drawbacks of the existing replacement methods, a direct application of them to the vehicles deployed by freight transportation companies remains uncertain.

14.6 Practical Challenges for the Fleet Replacement Problem

Typically fleet management for field services requires finding the right vehicle, of the right capacity, for the right business, and fitting the required features into the serviced work type. In practice, these decisions are twofold:

- First, decision aims at identifying the vehicles portfolio needs in terms of volume capacity, driving features (speed and driving wheels, for instance).
- Second, decision requires a system for calculating a replacement plan, from 1 to 5 years. This aims at ensuring the provision of the right brand, model, and vehicle asset supplier for each identified fleet item.

The second step can be modelled as a multi-objective combinatorial optimisation problem. However, there is not a single solution; as a matter of fact, the solution is in the form of a ranking of the technology and brands available based on the most economical and ecological choice. The accuracy of such a ranking is generally limited to a number of years; due to high variations in energy prices market, fleet managers generally are advised to plan 1 year in advance. Therefore, there is an important practical challenge: to increase the planning horizon to the full 4 years, taking into account all the uncertainties.

The combinatorial aspect of the operation is complicated by the fact that the matching of vehicle types and running technology depends both on the driver's behaviour and on the variation of usage over days, months, or years. For instance, a simple analysis suggests that the petrol engine tends to be cost effective when dealing with short annual mileage usage, and a mixed diesel and hybrid technology are suitable for normal distances while affording a risk exposure reduction. Moreover, the electric engine tends to be the optimal choice, from both risk and cost minimisation perspectives, when the annual mileage usage is high.

The following are some of the challenges faced by fleet provisioning:

- **The fleet provisioning needs to consider the mileage driven by the vehicles**. Thus, in the process of constructing a replacement tactical plan, we need to implement a method for forecasting annual mileage with a granularity at the vehicle type or service operations type level.
- **The length of equipment life is not fixed**. Even though the rental duration can be used as working hypothesis, in practice the replacement decision may happen before the planned end of life, depending on the maintenance cost, fuel prices variation forecast, electric energy recharge constraints, geography, and volume of the field service demand.
- **We need to find a balance between risk exposure and O&M cost minimisation**, taking into consideration the utilisation of vehicles and the frequency of long, medium, or short distance driven by each vehicle. A fine granularity analysis of mileage, fuel consumption, and geographical information

monitoring data will help in adjusting the approach for realising sustainable field operations.

– **There is a need to consider fuel price uncertainty**, the **variation of real fuel consumption** in each technology, leasing costs, and the accessibility of vehicles based on the data for accidents.

– **Robustness of the replacement plan.** If we consider a larger number of aspects in the model, then the analysis will be more accurate. If you want to introduce manufacturing costs into the model, you will require quote information from the enterprise processes; if you consider customer experience (service commitment delivered, number of visits before completing the task, asset missing, for instance), you will need to analyse the robustness of the replacement plan when environment or service engineering variables change. Furthermore, an analysis of the impact of the average speed of the vehicles on the fleet management decisions seems to be one of the other direction of research; however, this variable suffers generally from data quality issues, due to lack of links between tactical planning and the travel feedback from field workers: the use of an electronic box embedded in vehicles is an interesting alternative to improve the flow of information from operations to strategic planning, one of which should be considered if the improvements in fleet management outweigh the costs of installing and maintaining the system.

Additionally, the vehicle utilisation governance within a firm also has an important impact on fleet management. We can consider this issue if we analyse the fleet portfolio life cycle at an organisational level. In this framework, a vehicle is seen as an item that can be swapped across business units: in this case, the transfer of an unused vehicle from a line of business to another one would be a better alternative to rent a new vehicle. If we consider this new framework, several questions arise: Which option leads to the best cost risk and customer experience trade-off? How can the cost of vehicle reuse option be recorded?

This governance structure at a global level, when transforming the fleet portfolio and the impact on environment, requires support at a tactical level by:

• Planning the number of vehicles per technology (source of energy), capacity, and various mileages, in the short, medium, and long term
• Analysing risk exposure (taking into account the forecasted demand and supply life cycle)
• Considering the impact of such decisions on the customer experience

14.7   Conclusions

In this chapter, we provided a comprehensive literature review for different approaches regarding the asset replacement problem and its particular case of field service fleet. Specifically, if we consider the conventional vehicle replacement decisions that exist among fleet managers of the companies and the impact of emerging new technologies on adoption of optimal replacement policies, the main questions that should be addressed for the fleet manager are:

- First, what kind of vehicle technologies has a better performance in terms of cost efficiency?
- Second, what is the impact of market uncertainties on vehicle replacement decisions?
- Third, what are the best practices for replacing vehicles in the future?

The model suggested in Sect. 14.4.2 has the potential to address most of the drawbacks in the existing replacement methods. First, it takes into account the variability of vehicles' operational (running) costs. Indeed, the majority of the parameters of the model depend on time, and fixed and variable aspects are distinguished in cost parameters. In particular, the expected utilisation (annual mileage driven) per year is assumed as a variable in each year. In addition, CO_2 emissions costs are also taken into account.

Moreover, unlike most of the papers in the literature, the leasing option is considered as a way for financing the vehicles in the fleet system which is commonplace in the most of the commercial logistics systems. By taking into account leasing of the new vehicles at the beginning of each year for a finite time horizon (4–5 years), many issues regarding the optimal age (economic life) of vehicles and relation with age and utilisation will be resolved, due to young structure of the fleet system.

Nevertheless, the model assumes the availability of a certain number of historical inputs and of forecasted data such as fuel prices, fuel consumption, CO_2 prices and the utilisation trend of the vehicles along years. This data should be collected, updated, and processed with the application of a modern database. This database combined with the suggested model provides a decision support system for a strategic fleet management in any transportation company.

References

Bean JC, Lohmann JR, Smith RL (1984) A dynamic infinite horizon replacement economy decision model. Eng Economist 30(2):99–120

Bellman R (1955) Equipment replacement policy. J Soc Ind Appl Math 3(3):133–136

Chand S, McClurg T, Ward J (2000) A model for parallel machine replacement with capacity expansion. Eur J Oper Res 121(3):519–531

Chang C-C, Sheu S-H, Chen Y-L (2010) Optimal number of minimal repairs before replacement based on a cumulative repair-cost limit policy. Comput Ind Eng 59(4):603–610

Christer AH, Scarf PA (1994) A robust replacement model with applications to medical equipment. J Oper Res Soc 45(3):261–275

Dejax PJ, Crainic TG (1987) Survey Paper-A review of empty flows and fleet management models in freight transportation. Transport Sci 21(4):227–248

Dobbs IM (2004) Replacement investment: optimal economic life under uncertainty. J Bus Fin Account 31(5–6):729–757

Drinkwater RW, Hastings NAJ (1967) An economic replacement model. Oper Res Soc 18 (2):121–138

Eilon S, King JR, Hutchinson DE (1966) A study in equipment replacement. Oper Res Soc 17 (1):59–71

Feng W, Figliozzi M (2013) An economic and technological analysis of the key factors affecting the competitiveness of electric commercial vehicles: a case study from the USA market. Transport Res C Emerg Technol 26:135–145

Hartman JC (1999) A general procedure for incorporating asset utilization decisions into replacement analysis. Eng Economist 44(3):217–238

Hartman JC (2000) The parallel replacement problem with demand and capital budgeting constraints. Nav Res Logist 47(1):40–56

Hartman JC (2001) An economic replacement model with probabilistic asset utilization. IIE Transactions 33(9):717–727

Hartman JC (2004) Multiple asset replacement analysis under variable utilization and stochastic demand. Eur J Oper Res 159(1):145–165

Hartman JC (2005) A note on "a strategy for optimal equipment replacement. Prod Plann Cont 16 (7):733–739

Hartman JC, Lohmann JR (1997) Multiple options in parallel replacement analysis: buy, lease or rebuild. Eng Economist 42(3):223–247

Hartman JC, Murphy A (2006) Finite-horizon equipment replacement analysis. IIE Transactions 38(5):409–419

Jones PC, Zydiak JL, Hopp WJ (1991) Parallel machine replacement. Nav Res Logist 38 (3):351–365

Karabakal N, Lohmann JR, Bean JC (1994) Parallel replacement under capital rationing constraints. Manage Sci 40(3):305–319

Keles P, Hartman JC (2004) Case study: bus fleet replacement. Eng Economist 49(3):253–278

Kim HC, Ross MH, Keoleian GA (2004) Optimal fleet conversion policy from a life cycle perspective. Transport Res Transport Environ 9(3):229–249

Nakagawa T, Osaki S (1974) The optimum repair limit replacement policies. Oper Res Q 25 (2):311–317

Nakagawa T (1984) Optimal number of units for a parallel system. J Appl Probab 21:431–436

Oakford RV, Lohmann JR, Salazar A (1984) A dynamic replacement economy decision model. IIE Transactions 16(1):65–72

Rajagopalan S (1998) Capacity expansion and equipment replacement: a unified approach. Oper Res 46(6):846–857

Rajagopalan S, Singh MR, Morton TE (1998) Capacity expansion and replacement in growing markets with uncertain technological breakthroughs. Manag Sci 44(1):12–30

Redmer A (2009) Optimisation of the exploitation period of individual vehicles in freight transportation companies. Transport Res E Logist Transport Rev 45(6):978–987

Ritchken P, Wilson JG (1990) (m, T) group maintenance policies. Manag Sci 36(5):632–639

Simms BW, Lamarre BG, Jardine AKS, Boudreau A (1984) Optimal buy, operate and sell policies for fleets of vehicles. Eur J Oper Res 15(2):183–195

Suzuki Y, Pautsch GR (2005) A vehicle replacement policy for motor carriers in an unsteady economy. Transport Res A Pol Pract 39(5):463–480

Weissmann J, Jannini Weissmann A, Gona S (2003) Computerized equipment replacement Methodology. Transport Res Rec: Journal of the Transportation Research Board 1824:77–83

Yatsenko Y, Hritonenko N (2011) Economic life replacement under improving technology. Int J Prod Econ 133(2):596–602

Ye MH (1990) Optimal replacement policy with stochastic maintenance and operation costs. Eur J Oper Res 44(1):84–94

Part IV
Challenges, Outcomes and Future Directions

Chapter 15
Enabling Smart Logistics for Service Operations

Yingli Wang, Mohamed Naim, and Leighton Evans

Abstract Efficient and on-time execution of field tasks has been found to rely heavily on internal availability of inventories. However, the lack of flexibility in the way information flows along the logistics chain has led to poor inventory replenishment lead times. This results in delayed execution of field tasks and has a negative impact on customer experience. This chapter articulates the concept of communication flexibility, in the form of dimensions, by which operations managers may judge the ability of the logistics chain to configure and reconfigure information linkages in response to a changing environment. Until now the term 'communication flexibility' has been loosely used in the literature. This research establishes a more analytical definition that forms the foundation for more comprehensive empirical quantitative and qualitative research in the field of flexible operations. The research method is a combination of conceptual and literature review based research. The chapter proposes a conceptual model of intra-organisational communication flexibility which is composed of three levels, namely, transactional, operational and strategic. Each level consists of a number of dimensions and sub-dimensions that together define communication flexibility in logistics operations. Current research in the deployment of ICT in inventory projects is then considered in depth in order to preliminarily verify and validate the proposed model. This chapter provides an overview of current best practice and technological use in inventory management that emphasises the importance of visibility in the management of inventory achieved through ICT deployment.

Y. Wang (⊠) • M. Naim
Cardiff Business School, Cardiff, UK
e-mail: WangY14@cardiff.ac.uk

L. Evans
University of Swansea, Swansea, UK

G. Owusu et al. (eds.), *Transforming Field and Service Operations*,
DOI 10.1007/978-3-642-44970-3_15, © Springer-Verlag Berlin Heidelberg 2013

15.1   Introduction

The provision of logistics services, including inventory services, relies heavily on effective intra- and inter-organisational information exchange and communication, for which information and communications technology (ICT) is seen as a key enabler (Phillips and Wright 2009; Swafford et al. 2008). Effective inter-organisational communication helps to foster collaboration and reduce uncertainties and performance-related errors, enhancing operational efficiency (Holweg et al. 2005; Premkumar et al. 2005). Innovative logistics practices that are based on the use of ICTs are often referred to as 'smart logistics', which implies a flexibly and ability to cope with uncertainties (Uckelmann 2008).

The aim of this chapter is to develop and validate the concept of communication flexibility enabled by ICT in the context of inventory management and to identify potential and actual innovations in the field of smart logistics that can benefit inventory management, in particular with regard to visibility. Given that the use of ICT to manage enterprise-wide activities is relatively mature, this chapter will focus on the intra-organisational perspective where there are a variety of emerging technologies that have not been adequately researched in the literature. Intra-organisational communication flexibility is not well defined, but here it is defined as the extent to which organisations are able to configure and reconfigure their internal information linkages in response to changing environments and in particular how ICT can be deployed in a flexible manner to assess, facilitate and control the deployment and monitoring of inventory.

The rest of the chapter is organised as follows. In the next section, a systematic review of studies on flexibility and the enabling effect of ICT in the field of logistics is provided as a grounding and source of information for the application of current and future innovations to the field of inventory management. Next we propose a conceptual model of inter-organisational communication flexibility for inventory management. Following this, we detail current ICT deployments in the area of inventory management and explain their role in facilitating communication flexibility for inventory management in accordance with the conceptual model that has been developed. The final section draws conclusions and gives directions for future research.

15.2   Lesson from Logistics: Literature Review

Early data exchange systems, such as electronic data interchange (EDI), are often criticised for their adverse effects in facilitating flexible information transfer between and within organisations because they utilise rigid and complex interfaces and are costly to deploy (Badii and Sharif 2003). However, recent advances in ICTs, in particular web-based technologies, have altered the way information flows are managed and structured. Avoiding the problems inherent with costly and complex point-to-point integration of separate systems, web-based systems are designed for organisations and divisions to share a single platform (Christiaanse et al. 2004). Many third-party, cloud-based applications and software now exist to take advantage of web-based flexibility, such as TradeGecko, IBM's own mobile inventory management offerings or Bizelo. Such technological advances have made intra-organisational information connectivity more flexible and less costly, allowing increased visibility and opening up opportunities for better decision making, increased stability in operations and collaborative initiatives within organisations.

The idea that flexible connectivity automatically leads to increased communication flexibility and effectiveness is too simplistic in practice. In fact, in a recent global survey of 400 supply chain executives, 70 % still rate 'achieving the level of interaction and visibility they need' as a top challenge (Butner 2010), partly due to the increasing complexities for organisations in determining appropriate information linkages needed for successful operations. This need for visibility is directly applicable to the field of inventory management, where accurate locational and quantity data is critical to the accurate assessment of current inventories and forecasting. A 2-year study further confirms that there is a strong need from providers of logistics services to understand how to leverage the emerging technologies for competitive advantages (Huckridge et al. 2010). Therefore, there is a need to define and characterise communication flexibility, thus guiding practitioners to realise the potential practical utilisation of such a concept.

The concept of communication flexibility is of equally in need of development in academia. Much of the literature related to flexibility concentrates on manufacturing operations, including the early notable works of Slack (1987) and Gerwin (1987). More recently, the study of flexibility has extended from a manufacturing systems level to a supply chain level (Sanchez and Perez 2005).[1] Within the literature, there is limited evidence in explicitly defining and measuring ICT-enabled communication flexibility. Of those studies which directly relate to ICT-enabled logistics flexibility, it is found that they confirm the positive effect of ICT on flexibility but have limitations in providing in-depth insights into characterising communication flexibility and defining the phenomenon in its own right. For instance, Swafford et al. (2006, 2008) found that integrative ICT capability is a positive enabler of logistics and distribution flexibility in terms of range (options for storing and delivering products) and adaptability (ability to exercise the different logistics options within the range). In addition, a firm's ability to deploy an ICT infrastructure in support of the organisation-wide use of market information in logistics services can provide comparative advantages (Davis and Golicic 2010). Such studies are indicative of positive uses of ICT in intra-organisation connectivity. A survey of 198 companies in Hong Kong identified that the use of ICT improves operational adaptability under environmental uncertainty and thus leads

---

[1] For a comprehensive review of manufacturing and supply chain flexibility, one can refer to the work of Stevenson and Spring (2007) and Bernardes and Hanna (2009).

to improved cost performance of logistics (Wong et al. 2009). The study did not explicitly discuss the specific technologies used by the surveyed companies and so is only of value in broad terms.

One of the few papers which specifically investigates the mediating effect of information connectivity (including information sharing and collaboration) between flexible logistics programmes and flexibility outcomes (responsiveness, delivery competence and asset productivity) was Closs and Swink's (2005) work. They found that increased information connectivity leads to improved asset productivity, but does not lead to increased responsiveness. As the survey used in this study was conducted in 1998, when web-based technologies for intra-organisational integration was not yet widespread (or in some cases technologically feasible), there is a need for such empirical results to be reinvestigated given the substantive advances in ICT.

Only two works have been identified that address communication flexibility in a logistics context in an explicit manner. First, Naim et al. (2006), working specifically on transport flexibility, define organisational flexibility into two types: internal flexibilities which describe system behaviour and external flexibilities which determine the actual or perceived performance of the system. The paper argues that communication flexibility is an internal flexibility and briefly defines it as 'the ability to manage a range of different information types'. Zhang et al. (2006) used an alternative term, 'spanning flexibility', to describe the ability of a firm to provide information across a supply chain. This research also proposes a definition of 'supply chain information dissemination flexibility': the ability of a firm to collect and disseminate quickly the various data needed along a supply chain to respond resourcefully to the customer needs. Both these works have a loose definition of communication flexibility, and neither attempts to characterise communication flexibility with regard to ICT. A recent literature review conducted by Marasco (2008) confirms the lack of research in addressing the effect that ICT-enabled communication has on interactions between parties in logistics services.

This chapter fills the gap in adequately defining communication flexibility by addressing what the key dimensions of communication flexibility are in practice using ICT. Dimensions of communication flexibility should be measurable: in manufacturing operations flexibility measurement is relatively mature, providing managers with the ability to assess and improve performance and to benchmark between different operating systems (Slack 2005; Stevenson and Spring 2007). Whilst there has been some research in related fields, such as measuring the flexibility of ICT infrastructure (Byrd & Turner 2000), there is still a need to address how communication flexibility be measured. This chapter proceeds through an analytical development of a conceptual model of the constructs that define intra-organisational communication flexibility. The constructs of each layer are developed through additional literature and are discussed in the next section.

Fig. 15.1 Conceptual
framework for
communication flexibility

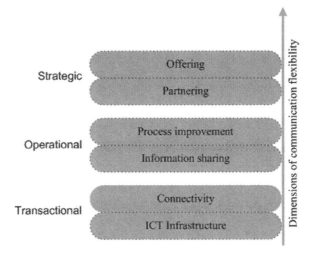

15.3   Inter-organisational Communication Flexibility

The model of communication flexibility proposed here is a three-layer model for intra-organisational communication flexibility as shown in Fig. 15.1. This conceptual model builds on the work of Ward and Peppard (2002) and Klein and Rai (2009) who argue that the impact of ICT to supply chain and inventory management can be classified into three levels:

- Strategic: an emphasis on planning and development of information provisions for strategic gains and competitive advantages
- Operational: an emphasis on the improvement of information flows within companies to unlock potential efficiency gains
- Transactional: an emphasis on automating data processing and enabling reliable data exchange within organisations

15.3.1   Transactional Layer

The transactional layer focuses on enabling effective and rapid data transfer (transactions) and is composed of two elements: *ICT infrastructure* and *connectivity*. ICT infrastructure here refers to hardware, software and networks. The strategic decisions from companies on how to use their ICT infrastructure have an obvious and significant impact on productivity and achieving competitive advantages, as ICT is a means of improving organisational productivity, enterprise innovation and service delivery (Mithas et al. 2011; Paulraj et al. 2008). Recent innovations in hardware utilisation have provided innovative ways in reducing the total cost of investment in ICT infrastructure. For example, cloud computing allows companies to avoid the cost of investment on an ICT infrastructure and the resulting costs incurred by the employment of maintenance and technical staff by renting the computing and storage capacity they require and paying a charge based on usage.[2] Sharif (2010) suggests that whilst cloud computing and SaaS are indeed innovations within ICT, the real innovation potential will emerge when these platforms allow new industries ways of doing business, connecting with and engaging with people. Another example derived from supply chain management is the increasing trend in the UK for small haulage companies to rent telematics kits for downloading digital tachygraphy data and real-time tracking of vehicles.[3] These approaches to ICT deployment reduce the rigidity and high fixed cost associated with planning, purchasing and maintaining of traditional ICT hardware solutions, increasing the ease of deployment across organisations and resulting in more flexibility in the infrastructure itself.

Software and computer code runs on hardware and executes the operational functions of ICT. Recent developments in software applications enable the possibility of greatly enhanced flexible provision of logistics. Specifically in the logistics field, around 20 years ago most carriers and shippers with their own fleets did their transport planning using spreadsheets, and many SMEs still do (Davies et al. 2007). There are now widely available software packages, such as computerised vehicle routing and scheduling (CVRS) or transport management systems (TMS) and warehouse management systems (WMS) that are commonly found in most organisations with transport logistics functions. These systems rapidly process information and produce the most economic routes and schedules for transport and warehouse management for the most cost-effective use of employee time and storage space. Particularly in an environment where inventory has a short shelf life, such systems are essential to aid the high-speed and efficient distribution.

Networks are easily defined as a series of computers connected via communication media (cabled or wireless) so as to transmit data (Rainer and Cegielski 2010). The two most common types of networks are local area networks (LANs) and wide area networks (WANs). Network systems provide critical links within organisations and make it possible for geographically distributed organisations/individuals to communicate and collaborate effectively. According to a survey by the Aberdeen Group (Viswanathan 2008), hosted networks in recent years have gained popularity, which again offer a flexible and cost-effective mechanism for deployment.

The second element of the transactional layer is *connectivity*. Connectivity refers to the ability of intra-organisational systems to gain access to information and support data exchange within organisations. Companies and individual users have to have quick access to information via ICT systems in responding to changes in the working environment. Prior to the widespread use of the Internet, many logistics

---

[2] Taking the example of Amazon, the company charges from \$0.095 to \$0.96 per hour on one server for computing capacity in the EU (Amazon.com 2011).

[3] One technology provider Quartix charges their client from £18.5 to £22.18 per month for leasing the telematics kit and accessing the tracking data via the Internet (Quartix.co.uk 2011).

information systems were internal systems that could only be accessed in fixed geographical locations. The pervasiveness of web technologies and wireless communication allows for access to information systems that transcends geographical proximity, resulting in communications from anywhere in the world if there is an Internet connection. Many organisations employ secure web-based interfaces that can be accessed remotely. Many organisations will grant their customers online access to their customer relationship management (CRM) module (normally as part of an ERP package). Third-party logistics companies like DHL offer shippers access to its internal business system to track the status of individual consignments via its online portal (DHL 2011). The food manufacturer Mars (UK) uses an Internet tracking system to connect all intra- and inter-logistics operations (The Institute of Grocery Distribution 2011). These solutions all increase the visibility of the supply chain and are illustrative of the operational benefits that can occur from effective data access and sharing principles.

Connectivity as a concept evokes links and linkages between nodes in a network, and the extent of these linkages is (according to Davis and Golicic 2010) known as the reach (i.e. e-connection with a wide audience) and range (share information across a variety of technological platforms). Semantic web services can provide an innovative solution to linkages and are being actively researched (Chituc et al. 2008; Panetto and Molina 2008).

15.3.2 Operational Layer

This second layer deals with two major issues: *information sharing* and *process improvement*.

Information sharing can be defined as the extent to which data is exchanged in real time and information asymmetries are reduced. Research on information sharing disproportionally focuses on information distortion and related costs and benefits (Folinas et al. 2004; Lee et al. 1997; Sahin and Robinson 2002; Samaddar et al. 2006). According to Donk (2008), there are few investigations about what is needed for information systems to provide seamless information flows (which would be of obvious benefit for communications within organisations). This chapter proposes that three constructs need to be considered to ensure information sharing: data quality, information visibility and speed of transactions. Data quality is often a primary concern for businesses, as the accuracy, consistency and completeness of data is critical to managerial decision making. In the past, many data errors occurred due to the manual inputting of data by operators. The use of automatic identification and data capture technologies can reduced data errors (Smith and Offodile 2002). Bar codes and radio frequency identification (RFID) are frequently used for inventory management in a variety of sectors (Jones 2011), providing automated and pre-encoded data. Voice picking is also seen as common in practice in retail sectors, with retailers like Tesco and ASDA. The use of real time satellite tracking of containers and trucks also provides timely, factual and accurate data for monitoring delivery performance (Wang and Potter 2007).

The increased complexity of supply chains and inventory tracking through globalisation has led to longer lead times for products, increased landed cost (the total cost of a product) and more goods-in-transit inventory. This necessitates expanded data exchanges and higher levels of connectivity to suppliers, carriers, customers and other parties like government bodies across multiple countries. This in turn has posed significant challenges to companies in obtaining the total pipeline visibility in order to respond quickly to inventory shortages as well as find ways to cut cost. Control and improvements can only be achieved if adequate visibility is provided.

"You cannot execute or optimize (processes) in an agile and responsive fashion in the absence of clear visibility." (Heaney 2009)

Many ICT visibility tools have been developed in practice in recent years to address the multi-enterprise visibility issue. For instance, Microsoft, IBM, JDA, Info and GT Nexus, to name only a few, have developed supply chain visibility solutions. There are visibility modules in supply chain management suites, visibility systems from carriers or freight forwarders and hosted commercial platforms. With regard to multimodal transport chains, a growing willingness to utilise hosted commercial platforms has been observed in shippers, in order to reduce the cost and complexity of maintaining such systems (Heaney and Sadlovska 2010).

Speed is concerned with how quickly visibility can be obtained. Visibility could realise within minutes, hours, days or even months. For instance, Tesco uses satellite tracking systems to monitor deliveries to their stores. The status of each individual delivery is fed into the transport management system every 5 min (Isotrack.com 2011). The store expecting the delivery will receive an automatic alert in the form of a geo-fence notification, where an RFID tag on the vehicle informs the system in the store that it is a specified distance from the completion of the vehicle's journey. Automation, integration and synchronisation in this manner are ICT-enabled ways to achieve speedy data transactions and visibility. The application of this kind of information sharing makes possible the second subcategory under operational layer: process improvement.

Most ICT tools are, ontologically, designed and developed for process improvements, particularly the streamlining of business processes such as inventory management (Turban and Volonino 2010). Some will also aid optimisation by enabling 'what-if?' scenario analysis. The origins of the use of ICT in logistics can be traced to the 1960s, when applications like inventory management systems, scheduling and billing systems were developed and first deployed by major organisations. These enterprise systems were stand-alone—automating single processes rather than entire operations. Material requirements planning (MRP) evolved in the 1970s from these stand-alone systems in order to integrate production, purchasing and inventory management functions. The 1980s saw the emergence of manufacturing resource planning (MRP II), which added labour and financial requirements into the system. Enterprise resource planning (ERP) in the 1990s is a logical and evolutionary development from these predecessor systems, with the aim of integrating all the transaction processing activities of all functional areas in the entire enterprise. These innovations in process improvements are built on a foundation of information sharing, as the efficacy of systems is contingent upon not only effectively communicated information but also accurate and timely information for maximum efficiency.

15.3.3 Strategic Layer

A primary motivation to deploy ICT systems in organisations is to utilise the developments in ICT for strategic gains. This third layer of communication flexibility deals with *partnering* and *offering*. Partnering refers to the ability to build and alter information linkages in response to changes in the business environment. This allows companies to be able to configure and reconfigure their supply chain structures in order to be responsive to customers' changing needs and emerging uncertainties.

Prior to the emergence of web-based technologies in the late 1990s, e-business integration between organisations was usually achieved through building dedicated linkages, for example, using enterprise application integration (EAI). This method of connectivity involved significant capital investment, long deployment time and high switching costs, and because of these factors, the systems were primarily utilised by large companies that could afford the development times and costs. The benefits of such systems were tangible however: once the system was set in place, a strong bond was created among linked organisations (Gosain et al. 2004). As a result of increasing business dynamics since the 1990s, organisations require more robust and reconfigurable linkages to deal with changes in the business environment, particularly in industries where product life cycles are short or where the business environment is dynamic, for example, the electronic sector. Hence, when there is a need for structural partnering change, the traditional methods of integration have been assessed as too rigid to meet such requirements (Edwards et al. 2001) as companies require the flexibility to configure or reconfigure their information linkages to reflect changing business arrangements and practices, which highly rigid systems do not allow.

Web-based systems make affordances for the possibility of flexible integration. Eschewing the costly and overly complex point-to-point integration network that is indicative of separate systems based in geographically or organisationally disparate locations, web-based systems are designed for participants to share a single system. A key emergent aspect of the rapid development of web technologies has been 'cloud computing' (Hayes 2008). Unlike traditional applications that are paid for with an upfront licence fee and installed on a company's own premises and network infrastructure, cloud computing systems are hosted by the vendor and paid for on a subscription basis. Cloud computing is often referred to as 'software as a service (SaaS)' (O'Sullivan 2007). These systems offer greater flexibility for collaboration within and between organisations and enable not only large companies but also small- and medium-sized companies to be able to use web-based technologies through flexible and affordable cost structures based on usage rather than development and licensing costs. Such technological advances serve as a catalyst to the development of new e-business logistics models. In an inter-organisational context, the adoption of such systems necessarily progresses with and requires inter-organisational trust and positive working relationships. In an intra-organisational context, one may think that these factors would be a given; even in cases where considerable intra-organisational rivalries exist and geographical or cultural differences prevent the fostering of positive intra-organisational cultures towards sharing, the tangible benefits of partnering of this kind should be catalytic in the adoption of these techniques if they fit the purpose and needs of the organisation and business environment.

Adapted from Gosain et al. (2004) and Sambamurthy et al. (2003), offering refers to the ability of organisational informational linkages to support changes in product or service offerings to customers. Limited lifecycles of products and services and the variability of customer demands are the driving forces for this attribute of flexibility. A commonly cited example from the 2000s is Dell, who allowed its customers to build their PCs to their own specifications by providing modular choices via a web-based interface. The interface was linked to an online order capturing system, and this was further linked with their in-house ERP system. The approach increased Dell's offering to customers in terms of variety and customisation but without a corresponding increase in cost. Mass customisation of this kind is not feasible without the supporting ICT technologies. Holmström et al. (2010) argue that by combining IT and rapid manufacturing technologies, original equipment manufacturers (OEMs) in an aircraft spare parts supply chain can meet the dual challenges of providing a variety of spare parts to their customers without holding excessive stock. The strategic decision to offer flexibility to customers through the use of ICT necessarily involves a robust and responsive ICT network and series of systems to create and transfer accurate and executable data to all relevant parts of the organisation and as such requires strategic decision taking in applying trust and faith in the ICT systems provided.

15.4 Current Developments in ICT Use in Inventory Management

Inventory management is already heavily reliant upon ICT and ICT-enabled processes, with many of the technologies mentioned in the above section (for instance, warehouse management software) deployed extensively across industries as a means to managing inventory levels and distribution channels. The following research is intended to highlight new innovations being deployed in inventory management and across a number of business sectors: telecommunications, health, utilities, construction, transport infrastructure and the automotive industry. This is done with the intention of illustrating current and potential uses of ICT to facilitate and realise communication flexibility at the levels outlined above and how inventory management can be improved with the use of existing ICTs that improve upon current deployments.

Recent research emphasises the importance and role of visibility (as a function of data) as a key enabler in improved inventory management. Alfaro and Rabade (2009) indicate from research in the grocery and fast-moving consumer goods sector that increased visibility brings both quantitative and qualitative benefits to firms, with the qualitative benefits being focussed on better working conditions and staff approval of operations following improved visibility through the use of vehicle tracking and product identification technology. The implementation of improvements at a transactional level identified in this research effect improvements at the organisational level of communication flexibility, in line with the topology in the theoretical model proposed. Becker (2011) argues that it is necessary that a company's processes are transparent (through the provision of data) if effective process management is to be realised, but there is a lack of effective and comprehensible methods for implementing this requirement at present, and, without the establishment of coherent and established methods, ad hoc solutions that do not leverage the possible advantages and efficiencies from the technology could be lost. De Brito and Van der Laan (2009) add that imperfect data has significant effects on inventory control, particularly with regard to reverse logistics and product returns, which take considerable resources to correct and amend data in the inventory management process. Berghout et al. (2011) add that although organisations are committed to improving ICT, they squander many opportunities to do so through inconsistencies in cost/benefit management practices that do not integrate operational benefits into investment analyses, a view shared by Haider et al. (2006) who emphasise that the stochastic nature of the variables involved makes such decision making problematic but these problems can be solved with effective and responsive measurement mechanisms for process improvements. Hertwig (2012) argues from case study data that the adoption of e-business solutions in the field of inventory management is still largely driven by external demands from other customers, bandwagon effects and herding behaviour, and less so the expectations and interests of professionals within a company, and that this model of demand can lead to suboptimal solutions and in extreme examples multiple systems with a lack of interoperability and data exchange being deployed within organisations, which increases rather than decreases efficiency and visibility.

The prevailing view from recent research is that increasing visibility of inventory is critical to improving operational performance. Emmanouilidis et al. (2009) offer the view that 24/7 connectivity (of networks and items, enabled through innovative technological use of systems such as RFID and the scanner networks and databases) for active data analysis, ready access to knowledge and information and growth of information networks are essential to the realisation of solutions in asset and inventory management, a view furthered by Carlos and Vieira (2009) who present the need for this kind of data processing environment as a main challenge facing organisations who are looking to develop agile working environments. Gospill et al. (2011) emphasise that such improvements in agility that necessitate accurate and timely data on inventory are dependent upon a fundamental understanding of data and that current systems and strategic thinking in organisations are both a barrier to realisation of the advantages of ICT through a fundamental lack of acknowledgement of the importance of data. Recent research by Shen et al. (2012) advocated agent-based web service solutions that allow for loosely integrated linkages as the optimal method for the capture of the necessary data for effective inventory management in construction projects where multiple agents are involved, although such an approach may be unnecessary where concerns over inventory management are within one organisation. In general, barriers to effective inventory management using ICT are currently identified as a lack of strategic willingness and knowledge in the employment of appropriate systems, difficulties with accurate and timely data capture and a lack of transparency and visibility of data on inventory. Improvements in these areas should, according to the research, result in improvements in inventory management.

15.4.1 Telecommunications

At present, there is scant recent research into inventory management in the telecommunications industry, which indicates both a gap in the literature and a need to address the industry with some urgency, as the industry itself meets the demands of rolling out net mobile networks and the maintenance of mature networks that require adequate inventory levels and an appreciation of what equipment is currently deployed in the vast networks that telecommunications companies establish and manage. Ala-Risku et al. (2010) provide the only recent research on inventory management in the telecommunications industry, analysing the effect of technology that tracks site installations and inventory have on the supply chain for a telecommunications company. The researchers note that as the scale and complexity of a major infrastructure project (the deployment of equipment in a new mobile network) increases, the role of accurate and robust tracking of installation work and inventory increases. Alignments between parties in the supply chain are more likely to break down with the absence of reliable inventory tracking; the researchers proposed that the introduction of inventory tracking technology would quickly improve performance with regard to project delivery. This finding has not been validated however, whether in practice or through further research. The proposed solution clearly looks to improve communication of data at both the transactional and operational levels of communication flexibility, but without confirmatory evidence, this research remains a guideline only.

15.4.2   Health

The health sector represents an analogous industry to the telecommunications industry, in that it is another industry where the tracking of and accounting for potentially valuable (and in this case life-saving) stock units is of critical importance. Baffo et al. (2009) argue that effective inventory management in this sector is critical as the costs of inventory and the potential savings (at a time of increased attention on economising and sourcing, potential savings within organisations is important) for both private and public healthcare systems become increasingly salient. The researchers propose an operational model (enabled by effective and accurate information sharing) where departments, wards and personnel are linked for the purpose of equipment and drug management, eliminating duplication and unnecessary costs.

Ting et al. (2011) report on the use of RFID technology in an attempt to realise this kind of efficiency in managing and tracking equipment and other inventory in the sector. The researchers utilised an exploratory case study method to implement and deploy a RFID system in a medical organisation and found that if due consideration was taken in the preparation, implementation and maintenance of the system, then the increased visibility of inventory realised by the deployment could realise efficiencies in organisational practice, illustrating how changes at a transactional level can benefit organisational practices in inventory management. Replication of these findings would be required for validation given the case study technique used, although the practical implementation of the technology in this case is encouraging as it moves beyond theoretical and nonevidence-based enthusiasm for RFID in inventory management. De Vries (2011) identifies the presence of multiple stakeholders in healthcare settings that may limit the enthusiasm for and understanding of appropriate ICT systems for the management of inventory, a finding that could be applicable to other industrial sectors.

15.4.3   Utilities

The term utilities are here used to group power (electricity), water and gas suppliers. Research in this area is of particular interest as these organisations typically have to deal with major, fixed inventories around installations that typically have lengthy product life cycles but which require consistent and constant maintenance due to their strategic and operational importance in everyday life. For example, El-Thalji and Liyanage (2010) used a critical review method to assess the state of inventory management in onshore and offshore wind farms (increasingly prevalent as environmental concerns become a major political issue). The issues with offshore wind farms are particularly daunting: design of electrical infrastructure, structural design, material choices for the maritime environment, site assessments, substructural design and maintenance, installation methods, supply chain management and logistics and technical service access are all issues that impact on the design of the facility and the management of inventory for the maintenance of the facility. The chapter advocates intelligent remote diagnostics (integrated with inventory management and operational systems) can facilitate the optimal management of such facilities.

Other research has looked at the domestic power market and the deployment of inventory in domestic and commercial applications. Like other sectors, the utilities industry needs solutions to tracking inventory that is deployed in the field and has considerable asset value for organisations. Mason et al. (2010) found that the deployment of wireless sensors and sensor networks can improve efficiency in the monitoring, maintenance and security of gas canisters (which have considerable scrap value). The tracking of these items realised financial and material gains and allowed for streamlining of maintenance and other operations. Research that has assessed how ICT can improve operational efficiencies in the power industry also includes Taylor et al. (2011), who detail how ICTs could enable real-time state estimation through linkages with smart metres and other deployed infrastructure in the domestic power market.

The generation of vast amounts of data resulting from the deployment of devices that are enabled to produce data will require novel ICT deployments that are able to enable near real-time state estimation for the most effective generation of power and use of resources across the power network. This kind of ICT usage, with an emphasis on utilisation and deployment on a mass scale at the transactional level of communication flexibility to inform the operational aspects of the industry, is indicative of a transferrable deployment of ICT that could benefit other sectors (such as the health or telecommunications industries). Huet et al. (2010) concur that the deployment of smart technologies in the network will realise operational benefits but draw attention to the fact that such smart grids are in themselves a source of stress for networks (particularly those characterised by aging assets) and that this is a potential concern for the industry as a whole.

Kangilaski (2009) approaches inventory management in the power sector from an alternative perspective, arguing in the context of European Union member states that as the European market should run as a continuous market, there should be no restrictions according to individual state borders. To prepare for this (which implicitly argues that this is not the case at present), asset and inventory management software must be deployed to monitor pan-European assets and inventory. The research concentrated on the Estonian power utility provider Eesti Energia and their deployment of distribution and transmission monitoring technologies. In an industry where mergers and acquisitions concerning cross-border organisations and organisations with a pre-existing cross-border profile are already a prominent feature, this kind of ICT deployment should be of great strategic benefit as well as providing immediate inventory management benefits if deployed effectively.

15.4.4 Transport

The techniques and ICT deployments in the maintenance and tracking of inventory in road and rail networks are, again, illustrative of attempts to bring cost rationalisations to large inventories with considerable deployment issues. Leijten and Koppenjan (2010) assessed asset management in the Dutch railway infrastructure and found that whilst ICT deployment strongly improved the efficiency of the company in both financial management and capacity, there was a resulting burden on asset managers that had not been anticipated. The deployment of ICT in this case resulted in more work and pressure to realise efficiencies than previously. However, little other research is currently available to validate these findings further, indicating that transport infrastructures are another area in need of research focus in inventory management and ICT.

15.4.5 Automotive

Most relevant research has derived from studies in the automotive industry, but whilst the above areas deal with considerable fixed inventories as well as consumables, inventory management in the automotive industry is focussed on consumables and spare parts. Hellstrom and Wiberg (2009) report on an empirical study within a production and assembly plant where the use of RFID-insured inventory inaccuracy was kept at a minimum, allowing for the improvement of factory processes through the realisation of visibility of inventory by improving transactional creation and exchange of data and analytic gains. Liu and Sun (2011) also report on information flow management with automotive parts, describing how utilising the 'Internet of things' concept (uniquely identifiable objects with embedded smart technology that produces data on tracking in an Internet-based information capture system) with parts allows for the emergence of vendor-managed inventory which can realise improvements and efficiencies in supply chain performance and management.

Lou et al. (2011) identify that currently a gap exists in the material and information flows within a consumables supply chain. This is because the information flows do not always reflect the material flows in real-time (or in a timescale that is commensurate with optimal inventory management beyond real time), but the techniques of RFID tagging and other computational embedding can help close this temporal gap and realise real-time information flows that are in accordance with material flows. Again, this research in the automotive sector emphasises the critical importance of visibility of inventory through the utilisation of smart technologies and the effective use of that information operationally and strategically.

15.5 Conclusions

The brief review of current literature across industrial sectors illustrates that current research and innovations in inventory management are concentrated around visibility of inventory, through the use of smart technologies such as RFID, to improve the accuracy of inventory management and realise operational gains through this increased accuracy. This approach to the improvement of inventory management through the use of ICT is, in the context of communication flexibility as outlined in this chapter, an attempt to improve the transactional level of communication flexibility in order to produce organisational and strategic benefits. As such, the improvement in flexibility at the transactional level through the provision of visibility through data improves communication flexibility by improving the basis for communication itself—information or data. In these recent research examples, the provision of item- and stock-specific data provides the basis for communication itself and allows for greater flexibility in communication through the provision of the data which is the content of communication.

The additional communication afforded by the provision of real-time or locational data facilitates changes in operational procedures and cultures at the operational level, and both influences strategic decision making and is in itself a product of strategic offering and strategic decision making with the aim to improve accuracy of inventory measurement and therefore identify efficiencies and areas for further efficiency. In this sense, the three levels of communication flexibility operate as a topological system (i.e. a layered and connected system) in which improvements or changes in the transactional level are reflected by changes in the operational and strategic level. These resultant changes can be further reflected by additional changes at the transactional level, either in the provision of more data or different data to suit operational and strategic needs. The model of communication flexibility in practice offers a view of data provision, accumulation and analysis as a cybernetic system where feedback loops allow for the accumulation of data and the improvement or alteration of operations and strategic decision making based on continuous, updated and accurate data provision.

The model of communication flexibility both anticipates and explains how improvements in the provision of data on inventory (increases in transactions within communications networks) can both affect and improve communication flexibility at subsequent levels, which can result in operational and strategic gains whilst also affecting changes in practices which would require inspection and reflection at the individual case level. These effects are, however, not supported by enough literature, research and data from actual organisations at present; the research presented in this chapter tentatively supports the notion that improvements in data provision (resulting in improved visibility of inventory) as an improvement in communication flexibility can result in operational improvements in inventory management concretised as reduced costs and wastage and improved service and delivery times. The overall paucity of research in the area though demands more research to validate the findings presented and in particular more evidence based on economic costs/benefits for guiding investment in the technology that could enable these gains in communication flexibility and performance in inventory management is required.

References

Ala-Risku T, Collin J, Holmstr MJ, Vuorinen J-P (2010) Site inventory tracking in the project supply chain: problem description and solution proposal in a very large telecom project. Supply Chain Manag: An International Journal 15:252–260

Alfaro JA, Rabade LA (2009) Traceability as a strategic tool to improve inventory management: a case study in the food industry. Int J Prod Econ 118:104–110

Amazon.com (2011) Amazon elastic computer cloud (Amazon Ec2). http://aws.amazon.com/ec2/#pricing. Accessed 27 July 2011

Badii A, Sharif AM (2003) Integrating information and knowledge for enterprise innovation. Logist Inform Manag 16(2):145–155

Baffo I, Confessore G, Liotta G, Stecca G (2009) A cooperative model to improve hospital equipments and drugs management. In: Camarhina-Matos L, Paraskakis I, Afsarmanesh H (eds) Leveraging knowledge for innovation in collaborative networks. Springer, Boston

Becker J (2011) Information models for process management—new approaches to old challenges. In: Carugati A, Rossignoli C (eds) Emerging themes in information systems and organization studies. Physica-Verlag HD, Heidelberg

Bernardes ES, Hanna MD (2009) A theoretical review of flexibility, agility and responsiveness in the operations management literature. Int J Oper Prod Manag 29(1):30–53

Berghout E, Nijland M, Powell P (2011) Management of lifecycle costs and benefits: lessons from information systems practice. Comput Ind 62:755–764

Butner K (2010) The smarter supply chain of the future. Strat Leader 38(1):22–31

Byrd TA, Turner ED (2000) An exploratory analysis of the information technology infrastructure flexibility construct. J Manag Inf Syst 17(1):167–208

Carlos FP, Vieira NDJ (2009) Information technology and communication and best practices in it lifecycle management. J Technol Manag Innovat 3:80–94

Chituc CM, Toscano C, Azevedo A (2008) Interoperability in collaborative networks: independent and industry-specific initiatives—the case of the footwear industry. Comput Ind 59(7):741–757

Christiaanse E, Diepen TV, Damsgaard J (2004) Proprietary versus internet technologies and the adoption and impact of electronic marketplaces. J Strat Inform Syst 13(2):151–165

Closs DJ, Swink M (2005) The role of information connectivity in making flexible logistics programs successful. Int J Phys Distrib Logist Manag 35(4):259–277

Davies I, Mason R, Lalwani CS (2007) Assessing the impact of ICT on UK general haulage companies. Int J Prod Econ 106(1):12–27

Davis DF, Golicic SL (2010) Gaining comparative advantage in supply chain relationships: the mediating role of market-oriented IT competence. J Acad Market Sci 38(1):56–70

De Brito MP, Van der Laan EA (2009) Inventory control with product returns: the impact of imperfect information. Eur J Oper Res 194(1):85–101

De Vries J (2011) The shaping of inventory systems in health services: a stakeholder analysis. Int J Prod Econ 133(1):60–69

DHL (2011) Track DHL express shipments. http://www.dhl.co.uk/en/express/tracking.html, Accessed Feb 4 2011

Donk DPV (2008) Challenges in relating supply chain management and information and communication technology. Int J Oper Prod Manag 28(4):308–312

Edwards P, Peters M, Sharman G (2001) The effectiveness of information systems in supporting the extended supply chain. J Bus Logist 22(1):1–27

El-Thalji I, Liyanage JP (2010) Integrated asset management practices for Offshore wind power industry: a critical review and a road map to the future. School of Engineering, Linnaeus University, Sweden

Emmanouilidis C, Liyanage JP, Jantunen E (2009) Mobile solutions for engineering asset and maintenance management. J Qual Mainten Eng 15:92–105

Folinas D, Manthou V, Sigala M, Vlachopoulou M (2004) E-Volution of a supply chain: cases and best practices. Internet Res 14(4):274–283

Gerwin D (1987) An agenda for research on the flexibility of manufacturing processes. Int J Oper Prod Manag 7(1):38–49

Gosain S, Malhotra A, Sawy OAE (2004) Coordinating for flexibility in E-business supply chains. J Manag Inform Syst 21(3):7–45

Gospill J, McAlpine H, Hicks B (2011) Trends in technology and their possible implications on PLM: looking towards 2020. In: 8th international conference on project life cycle management (PLM 11). University of Eindhoven, Eindhoven

Haider A, Koronios A, Quirchmayr G (2006) You cannot manage what you cannot measure: an information systems based asset management perspective. In: Mathew J, Kennedy J, Ma L, Tan A, Anderson D (eds) Engineering asset management. Springer, London

Hayes B (2008) Cloud computing. Comm ACM 51(7):9–11

Heaney B (2009) Integrated transportation management: improve responsiveness with real-time control of execution. Aberdeen Group. http://www.aberdeen.com/aberdeen-library/6016/RA-integrated-transportation-management.aspx. Accessed 21 Jan 2010

Heaney B, Sadlovska V (2010) Supply chain visibility excellence. Aberdeen Group. http://www.aberdeen.com/Aberdeen-Library/6027/RA-supply-chain-visibility.aspx. Accessed 21 Jan

Hellstrom D, Wiberg M (2009) Exploring an open-loop RFID implementation in the automotive industry. IEEE conference on emerging technologies and factory automation (ETFA) 2009, 22–25 Sept 2009, pp 1–4

Hertwig M (2012) Institutional effects in the adoption of e-business-technology: evidence from the German automotive supplier industry. Inform Organ 22:252–272

Holmström J, Partanen J, Tuomi J, Walter M (2010) Rapid manufacturing in the spare parts supply chain. J Manuf Tech Manag 21(6):687–697

Holweg H, Disney SM, Holmström J, Småros J (2005) Supply chain collaboration: making sense of the strategy continuum. Eur Manag J 23(2):170–181

Huckridge J, Potter A, Wang Y, Beresford A, Naim M (2010) Enabling multimodal transport: a Welsh perspective. In: Proceedings of the 15th logistics research network conference, Harrogate, UK

Huet O, Guillaume C, Gaudin C (2010) Joint assets. IEEE Power Energy Magazine 8:88–93

Isotrack.com (2011) Tesco video. http://www.isotrak.com/casestudies/retail.php, Accessed 4 Feb 2011

Jones N (2011) Near field communications (NFC) – the next step after RFID. http://www.globallogisticsmedia.com/articles/view/near-field-communications-nfc–the-nextstep-after-rfid. Accessed 31 Jan 2011

Kangilaski T (2009) Asset management software implementation challenges for electricity companies. Industrial Electronics. IECON '09. 35th annual conference of IEEE, 3–5 Nov 2009, pp 3575–3580

Klein R, Rai A (2009) Interfirm strategic information flows in logistics supply chain relationships. MIS Quart 33(4):735–762

Lee HL, Padmanabhan V, Whang S (1997) Information distortion in a supply chain: the bullwhip effect. Manag Sci 43(4):546–558

Leijten M, Koppenjan JFM (2010) Asset management for the Dutch railway infrastructure. In: 2010 third international conference on Infrastructure systems and services: next generation Infrastructure systems for eco-cities (INFRA), 11–13 Nov, pp 1–6

Liu X, Sun Y (2011) Information flow management of vendor-managed inventory system in automobile parts inbound logistics based on internet of things. J Software 6:8

Lou P, Liu Q, Zhou Z, Huaiqing W (2011) Agile supply chain management over the internet of things. In: 2011 International conference on management and service science (MASS), 12–14 Aug 2011, pp 1–4

Marasco A (2008) Third-party logistics: a literature review. Int J Prod Econ 113(1):127–147

Mason A, Shaw A, Al-Shamma'a AI (2010) Inventory management in the packaged gas industry using wireless sensor networks. In: Mukhopadhyay SC, Leung H (eds) Advances in wireless sensors and sensor networks. Springer, Berlin, Heidelberg

Mithas S, Ramasubbu N, Sambamurthy V (2011) How information management capability influences firm performance. MIS Quart 35(1):237–256

Naim MM, Potter A, Mason R, Bateman N (2006) The role of transport flexibility in logistics provision. Int J Logist Manag 17(3):297–311

O'Sullivan D (2007) Software as a service: developments in supply chain it. Logist Transport Focus 9(3):30–33

Panetto H, Molina A (2008) Enterprise integration and interoperability in manufacturing systems: trends and issues. Comput Ind 59(7):641–646

Paulraj A, Lado AA, Chen IJ (2008) Inter-organizational communication as a relational competency: antecedents and performance outcomes in collaborative buyer–supplier relationships. J Operat Manag 26(1):45–64

Phillips PA, Wright C (2009) E-Business's impact on organizational flexibility. J Bus Res 62 (11):1071–1080

Premkumar G, Ramamurthy K, Saunders C (2005) Information processing view of organisations: an exploratory examination of fit in the context of interorganisational relationships. J Manag Inform Syst 22(1):257–294

Quartix.co.uk (2011) Vehicle tracking lease price. http://www.quartix.net/content/vehicle-tracking-lease.asp, Accessed 27 July 2011

Rainer RK, Cegielski CG (2010) Introduction to information systems: enabling and transforming business, 3rd edn. Wiley, New Jersey

Sahin F, Robinson EP (2002) Flow coordination and information sharing in supply chains: review, implications, and directions for future research. Decis Sci 33(4):505–536

Samaddar S, Nargundkar S, Daley M (2006) Inter-organizational information sharing: the role of supply network configuration and partner goal congruence. Eur J Operat Res 174(2):744–765

Sambamurthy V, Bharadwaj A, Grover V (2003) Shaping agility through digital options: reconceptualising the role of information technology in contemporary firms. MIS Quart 27 (2):237–263

Sanchez AM, Perez MP (2005) Supply chain flexibility and firm performance: a conceptual model and empirical study in the automotive industry. Int J Operat Prod Manag 25(7):681–700

Sharif AM (2010) It's written in the cloud: the hype and promise of cloud computing. J Enterprise Inform Manag 23(2):131–134

Shen W, Hao Q, Xue Y (2012) A loosely coupled system integration approach for decision support in facility management and maintenance. Autom Construct 25:41–48

Slack N (1987) The flexibility of manufacturing systems. Int J Operat Prod Manag 7(4):35–45

Slack N (2005) The changing nature of operations flexibility. Int J Operat Prod Manag 25 (12):1201–1210

Smith AD, Offodile F (2002) Information management of automatic data capture: an overview of technical developments. Inform Manag Comput Secur 10(3):109–118

Stevenson M, Spring M (2007) Flexibility from a supply chain perspective: definition and review. Int J Operat Prod Manag 27(7):685–713

Swafford PM, Ghosh S, Murthy N (2006) A framework for assessing value chain agility. Int J Operat Prod Manag 26(2):118–140

Swafford PM, Ghosh S, Murthy N (2008) Achieving supply chain agility through it integration and flexibility. Int J Prod Econ 116(2):288–297

Taylor GA, Wallom DCH, Grenard S, Yunta Huete A, Axon CJ (2011) Recent developments towards novel high performance computing and communications solutions for smart distribution network operation. In: 2nd IEEE PES international conference and exhibition on Innovative Smart Grid Technologies (ISGT Europe), 5–7 Dec 2011, pp 1–8

The Institute of Grocery Distribution (2011) Guide to transport technology case study: Mars— Supply chain connectivity. http://www.igd.com/index.asp?id=1&fid=5&sid=43&tid=59& foid=52&cid=1166. Accessed 6 Feb 2011

Ting S, Kwok S, Tsang A, Lee W (2011) Critical elements and lessons learnt from the imple- mentation of an RFID-enabled healthcare management system in a medical organization. J Med Syst 35:657–669

Turban E, Volonino L (2010) Information technology for management: transforming organisa- tions in the digital economy, 7th edn. Wiley, New Jersey

Uckelmann D (2008) A definition approach to smart logistics. In: Balandin S, Moltchanov D, Koucheryavy Y (eds) Next generation teletraffic and wired/wireless advanced networking. Springer, Berlin, Heidelberg, pp 273–284

Viswanathan N (2008) Process collaboration in multi-enterprise supply chains – leveraging the global business network. Aberdeen Research

Wang Y, Potter A (2007) The application of real time tracking technologies in freight transport. The IEEE third international conference on signal-image technology and Internet-based systems, Shanghai, China, pp 298–304

Ward J, Peppard J (2002) Strategic planning for information systems, 3rd edn. Wiley, New York

Wong CWY, Lai KH, Ngai EWT (2009) The role of supplier operational adaptation on the performance of it-enabled transport logistics under environmental uncertainty. Int J Prod Econ 122(1):47–55

Zhang Q, Vonderembse MA, Lim JS (2006) Spanning flexibility: supply chain information dissemination drives strategy development and customer satisfaction. Supply Chain Manag 11(5):390–399

Chapter 16
Measuring and Managing the Benefits from IT Projects: A Review and Research Agenda

Crispin R. Coombs, Neil F. Doherty, and Irina Neaga

Abstract There is growing agreement that organisations must explicitly plan for and proactively manage the realisation of benefits, if a new technology is to deliver real value to its host organisation. In particular, benefits need to be leveraged through carefully planned and co-ordinated programmes of organisational change and ongoing organisational adaptation. Inevitably these insights have encouraged academics, consultants and practitioners to develop tools and techniques that explicitly support the benefits realisation process. Unfortunately, even when organisations have adopted such prescriptions, tools or panaceas, the outcome from software projects still often disappoints users and managers alike. Based upon a thorough review of the existing literature, we begin by critically evaluating the benefits management literature and argue that before organisations can meaningfully manage benefits, they must be able to effectively measure benefits. We then critique the existing benefits measurement literature to assess whether the current measurement tools are sufficiently robust and effective, to facilitate benefits management approaches. The chapter concludes by proposing an agenda that identifies the many areas in which future research projects could be fruitfully conducted.

16.1 Introduction

'If you can't measure it, you can't manage it'. (Thorp 1998)

IT-related organisational transformation can engender changes in business processes and work practices, which may ultimately deliver value by reducing costs, increasing output quality, enabling new product development or improving

C.R. Coombs (✉) • N.F. Doherty
Centre for Information Management, Loughborough University, Leicestershire, UK
e-mail: C.R.Coombs@lboro.ac.uk

I. Neaga
Faculty of Business, Plymouth University, Plymouth, Devon, PL4 0LJ

G. Owusu et al. (eds.), *Transforming Field and Service Operations*,
DOI 10.1007/978-3-642-44970-3_16, © Springer-Verlag Berlin Heidelberg 2013

customer service (Brynjolfsson and Hitt 2000). Indeed, Gregor et al. (2006) found that variations in the level of organisational change can explain the differential effects of computer use on productivity across organisations. The recognition that most benefits from IS/IT come from changes in the way an organisation does business and not from the introduction of the new technology itself (Marchand and Peppard 2008) has engendered a whole new field of management practice and enquiry: *benefits management*. Benefits management (BM) has been defined as '*the process of organizing and managing, such that the potential benefits arising from the use of IT are actually realized*' (Ward and Elvin 1999). A number of previous studies have attempted to highlight the need for proactive management of organisational change in formal benefits realisation approaches to improve the outcomes of systems development projects (Remenyi et al. 1997; Changchit et al. 1998). Indeed, Doherty et al. (2012) consider the instigation of an organisational change process that complements the new information system's functionality as one of the defining characteristics of the benefits realisation approach. However, it is difficult if not impossible to proactively manage the process of benefits realisation management without effective tools to measure and monitor benefits.

Against this background, the primary aim of the current study was to review the academic literature that investigates different aspects of benefits measurement in order to identify and critique the benefit measurement approaches currently available. In so doing, we wanted to understand the extent to which current measurement techniques are *fit for purpose* in terms of facilitating the effective deployment of benefits realisation management approaches. Moreover, we were keen to identify the key priorities for future research regarding IS-/IT-enabled benefit measurement and management. The structure of this chapter is as follows: The next section provides a conceptualisation of IS-/IT-enabled benefits and discusses possible benefit classification schemes. The process of identifying and reviewing key contributions to the benefits measurement literature is then explained, before an overview of the main themes emerging from this review is presented. The chapter concludes by proposing an agenda that identifies the many areas in which future research projects could be fruitfully conducted.

16.2   Conceptualising IS-/IT-Enabled Benefits

Before we can discuss how benefits might best be measured, it is important to review how the term *IS-/IT-enabled benefit* has been conceptualised in the extant literature. This task is not as easy as it may sound, because there are few clear definitions, and many studies simply identify examples of the types of benefit that IS/IT investments deliver or classifications of these benefit types (e.g. Farbey et al. 1993; Giaglis et al. 1999; Shang and Seddon 2002), rather than attempting to explicitly defining this term. For example, Sanchez and Robert (2010) argue that all benefits can be classified as tangible or intangible depending on the ease with which they can be quantified. Remenyi et al. (1993) have defined '*a tangible benefit as one which directly affects the firm's profitability'*. The word *directly*, in this definition, attempts to draw a distinct line between tangible and intangible benefits. For example, it is fairly clear that an IT system contributing to a direct cost reduction is more tangible than one that is designed to deliver management information, which may, over time, facilitate improvements in organisational decision-making. However, Murphy and Simon (2002) argue that the delineation between tangible and intangible is often quite fuzzy. In this context, quantifiable benefits are differentiated from tangible benefits in that quantifiable benefits may be measured easily but may, or may not, directly affect a firm's profitability.

Some authors have attempted to devise benefit classifications that attempt to further decompose the broad categories of tangible/intangible or direct/indirect. For example, Williams and Parr (2006) have decomposed tangible benefits into financial and non-financial benefits. Giaglis et al. (1999) attempt to sidestep the issue of the degree of alignment between benefit tangibility and benefit directness by presenting a matrix of IS benefit types. In their matrix they differentiate between the following four benefits types: hard benefits (e.g. cost reduction), intangible benefits (e.g. improved decision-making), indirect benefits (e.g. the implementation of a LAN allowing future systems to be implemented if required) and strategic benefits (e.g. a new business strategy or better market positioning of the firm).

Strategic benefits have also been included in two other benefit classification schemes. Farbey et al. (1993) include strategic benefits as one of the five possible categories of generic benefits from IS/IT that they constructed from empirical study of project evaluation in 16 organisations. They also include management benefits (e.g. flatter organisational structure), operational benefits (e.g. timeliness and accessibility of data), functional benefits (e.g. enforcement of regulatory or legal requirements) and support benefits (e.g. improved recruitment and retention processes). These five categories are based on Mintzberg's (1983) view of the structure of an organisation. Ward and Daniel (2006) comment that whilst the first three categories remain relatively easy to understand, the last two categories are difficult to distinguish reducing the effectiveness of the framework.

A further classification is proposed by Shang and Seddon (2002). Their framework is based on the potential benefits available from ERP system implementations and has some similarities to Farbey et al.'s classification also including strategic and operational dimensions. The first three categories of Shang and Seddon's (2002) framework are based on Anthony's (1965) classification of management activity into three levels: operational, managerial and strategic. To these three levels they add IT infrastructure and organisational benefits. Shang and Seddon's framework has the advantage of being more clearly defined and derived from practitioner assessments of realised benefits, rather than planned or potential benefits. They also start to consider which types of measure may be more appropriate for measuring different benefit types differentiating between tangible and intangible benefit measures.

Whilst classifications are useful in that they provide useful insights into the types of benefits that may be realised from IT implementations, it is often difficult to apply them in practice, as the boundaries between categories are often indistinct.

Against this backdrop, other academics have attempted to define benefits, rather than categorise them. For example, Thorp (1998) defines a benefit as *'an outcome whose nature and value are considered advantageous by an organization'*. Bradley (2006, p18) provides a similar perspective defining a benefit as *'an outcome of change which is perceived as positive by a stakeholder'*, and Ward and Daniel (2006) continue the emphasis on stakeholders, by defining benefits as *'an advantage on behalf of a particular stakeholder or group of stakeholders'*. Sanchez and Robert (2010) provide a greater emphasis on the measurement aspect of benefits in their definition, stating that *'benefits are measurable improvements perceived to be a value by one or more of the stakeholders'*.

Conceptualisations and definitions of benefits both provide useful academic insights, but they say very little about how benefits are actually measured, in practice. Consequently we initiated a systematic literature review to investigate and critique the benefits measures that had been reported in the academic literature.

16.3   The Academic Literature on Benefits Measurement

Having reviewed some of the ways in which the benefits of IS/IT have been conceptualised in the literature, it is important to turn our attention to the rather more practical issue of what we know about the reality of measuring benefits. This chapter draws on a sample of references drawn from the literature on benefit measurement and management. Due to the multidisciplinary nature of the interest in the benefits from IT projects, we searched Web of Science, ABI/Inform, Business Source Premier, EI Compendex (Engineering Village), Scopus, Google Scholar, Inspec and IEEE electronic databases for English language journal peer-reviewed articles that were published between January 1990 and June 2012. We searched for articles, using a range of terms that included *benefits realization, benefits management, business benefit, benefits evaluation, business value, value engineering, value management* or *value realization* and *measurement, measures, assessment* or *valuation* in their title, abstracts or keywords. We combined the above search terms using lemmatisation, where available, and wildcard and truncation to pick up alternatives and to account for UK and US spellings. This search generated about 100 articles, which were imported into RefWorks. The titles and abstracts of these articles were examined to classify the literature and we ultimately identified 27 articles that focused on measuring benefits or IS/IT, which we now use as the focal point for the remainder of this chapter. Initial reading and forward and back citation analysis of key articles identified several additional papers that were included in the review resulting in a total of 32 articles for analysis (see Table 16.1).

An initial observation from the list is that the number of relevant articles, identified over this 20 year a time period, is relatively low, and it highlights the extent to which the concept of IS-/IT-enabled benefits measurement has been neglected and remains underdeveloped. This exercise also provided some interesting insights into the research methods used to study benefits measurement. The

Table 16.1 Academic articles on measuring IS-/IT-enabled benefits 1990–2012

Author(s)	Research topic	Conceptual	Survey	Modeling	Case study	Action Research
Martin-Oliver & Salas-Fumas (2012)	Measuring IS/IT to business value relationship			✓		
Berghout et al. (2011)	Management of costs and benefits throughout IS/IT lifecycle for IS/IT benefits management				✓	
Nevo & Wade (2011)	IS/IT to business value relationship		✓			
Gorla et al. (2010)	Measuring IS/IT quality improvement		✓			
Sanchez & Robert (2010)	Measuring benefits at project portfolio level	✓				
Gacenga et al. (2010)	Adoption of benefits measurement techniques		✓			
Lin et al. (2010)	IS/IT to business value relationship			✓		
Lin (2009)	Measuring value of IS/IT at country level to investigate IS/IT to business value relationship			✓		
Argyropoulou et al. (2009)	Development of framework for IS/IT evaluation				✓	
Jeffers et al. (2008)	Measuring IS/IT to business value relationship		✓			
Fox (2008)	Measuring disbenefits reliability and utilization					✓
Rao et al. (2008)	Model for measuring information quality			✓		
Buccoliero et al. (2008)	Assesses costs and benefits from different stakeholder perspectives for IS/IT evaluation				✓	
Tallon & Kraemer (2007)	Use of manager perceptions as proxy measures for business benefit levels		✓			
Sharif & Irani (2006)	Modeling tangible/intangible aspects of IS/IT evaluation				✓	
Yu et al. (2006)	Benefits planning for IS/IT evaluation				✓	
Rehesaar & Mead (2005)	Extending benefit costs analysis for IS/IT investment justification	✓				
Love et al. (2005)	Measuring different types of IS/IT enabled benefits		✓			
Ray et al. (2004)	Measuring IS/IT to business value relationship		✓			
Kleist (2003)	Agenda for new existing measures of IS/IT projects	✓				
Bresnahan et al. (2002)	Measuring IS/IT to business value relationship			✓		
Mukhopadhyay & Kekre (2002)	Identifies benefit measures for IS/IT in supply chain			✓		
Hitt et al. (2002)	Measuring IS/IT to business value relationship			✓		
Murphy & Simon (2002)	Measuring intangible benefits for IS/IT evaluation				✓	
Skok et al. (2001)	Measures for IS/IT evaluation		✓			
Benaroch & Kauffman (1999)	Measuring IS/IT to business value relationship			✓		
Brynjolfsson & Hitt (1998)	Measuring IS/IT to business value relationship	✓				
Mirani & Lederer (1998)	Develops new framework for measures for organizational benefits		✓			
Klein et al. (1997)	Measures appropriate for different system types		✓			
Powell & Dent-Micallef (1997)	Measuring IS/IT to business value relationship		✓			
Barua et al. (1995)	Measuring IS/IT to business value relationship			✓		
Smith & McKeen (1993)	Measuring IS/IT to business value relationship	✓				
Totals		**5**	**11**	**9**	**6**	**1**

findings presented in Table 16.1 indicate that the most common method adopted, within our sample of papers, was the survey strategy, usually across multiple firms, to investigate one of the following:

- The relationship between IS/IT and the delivery of business value (e.g. Nevo and Wade 2010; Jeffers et al. 2008; Powell and Dent-Micallef 1997)
- The validity of an author-developed measure (e.g. Tallon and Kraemer 2007; Love et al. 2005; Mirani and Lederer 1998; Klein et al. 1997)
- The approaches taken to measure benefits by practitioners (e.g. Gacenga et al. 2010)

Author-developed measures were the most common focus of the conceptual papers (Sanchez and Robert 2010; Rehesaar and Mead 2005; Kleist 2003; Smith and McKeen 1993), and the relationship between IS/IT investment and the delivery of business value was the primary concern of papers that adopted a financial or economic modelling techniques. A small number of studies had adopted a case study or action research approach to investigate aspects of benefit measurement compared to those following a survey design. These studies tended to focus more on softer aspects of benefits measurement addressing topics such as:

- Intangible benefits (e.g. Sharif and Irani 2006; Murphy and Simon 2002)
- Disbenefits (e.g. Fox 2008)
- Different stakeholder perspectives on benefits assessment (e.g. Buccoliero et al. 2008)

In addition to shedding light on the different research approaches reported in the literature, our study also provides important new insights into the specific types of measure that can be adopted. The remainder of this section critically discusses the two most common types of measure, namely, objective and perceptual measures.

16.4   Objective Measures of IS-/IT-Enabled Benefits

Nevo and Wade (2010) argue that the business value of IT (BVIT) is central to the IS discipline. Many researchers have attempted to apply economic measures to assess whether the anticipated benefits of IS/IT investments are being realised. Such *objective* measures include profitability, productivity, costs, quality, operative efficiency, consumer surplus and Torbin's q (Lin 2009). Many researchers have attempted to investigate how such objective measures are affected by organisational IT adoption practices. For example, Jeffers et al. (2008) report that whilst some studies have shown a positive relationship between IT spending and firm performance (Hitt et al. 2002), others have reported more mixed or negative associations (Nevo and Wade 2010; Lin 2009). Researchers have responded to these mixed results in two ways. One stream of research has focused on building *process-oriented* models of value creation. These scholars argue that IS/IT investments can only be measured at the intermediate process level as it is at this level that

IT-enabled contributions can be seen (Barua and Mukhopadhyay 1995; Ray et al. 2004). A second stream of research considers organisational resources and investments complementary to IT (Powell and Dent-Micallef 1997; Brynjolfsson and Hitt 1998; Bresnahan et al. 2002). However, it has been argued that the application of objective, accounting-oriented measures has significant weaknesses. For example, Smith and McKeen (1993) argue that return on investment is not suitable for capturing intangible aspects that may be essential for firm survival in an industry. Revenue growth rates are attractive because systems that are designed to offer new products or services are designed to increase market share. However, isolating an IS/IT investment contribution to revenue growth is likely to be problematic making linking individual IS/IT project costs with subsequent benefits difficult. Smith and McKeen (1993) argue that there are similar problems of isolation when attempting to apply return on management or profits as a measure of business benefits. However, despite such measurement problems, organisations still seem focused on gathering objective measures even though the business reasons for investing in these systems are often to provide intangible benefits (Petter et al. 2012). Consequently, although some authors are retaining objective measures (Lin 2009; Martin-Oliver and Salas-Fumas 2012), others have moved to *perceptual* measures.

16.5 Perceptual Measures of IS-/IT-Enabled Benefits

Several researchers have attempted to move away from objective measures and offer alternative instruments for assessing benefits from IS/IT investments. These approaches mainly rely on perceptual measures of benefits and the quality of respondent's recall rather than objective measures. For example, Mirani and Lederer (1998) developed an instrument to measure three dimensions of organisational benefits: strategic, informational and transactional using a Likert scale ranging from not a benefit to very important. More recently, Jeffers et al. (2008) and Nevo and Wade (2010) have relied on survey harvests of self-reported practitioner data on various aspects of firm performance and IS/IT contribution. Tallon and Kraemer (2007) provide a useful contribution to this debate as they assessed the efficacy of executives' perceptions of IT impacts at both the process and firm levels. They compared data collected via a perceptual survey of 196 executives views on their firm's IS/IT investment and performance with financial data for the same firms obtained from Standard and Poor's Compustat database. They conclude that whilst perceptions were not a perfect proxy for hard-to-find objective measures, they were *'sufficiently accurate, credible, and unbiased as to constitute a viable approach to IT impact assessment'*.

An alternative approach to evaluating the performance of IS/IT and therefore the delivery of benefits is proposed by Skok et al. (2001) who advocate the use of *importance–performance (I–P) maps*. I–P maps are matrix-based techniques that seek to capture stakeholder perceptions of importance and performance, using

Likert or numerical scales, which make them easy to analyse and interpret. Skok et al. (2001) assessed the application of this technique within a case study organisation and found it to be simple to administer and interpret, although it sacrifices depth for breadth of coverage. An alternative approach to addressing the challenge of assessing the value of IS/IT investments in a context of human, organisational, social and technical complexities is provided by Sharif and Irani (2006). They apply fuzzy cognitive mapping (FCM) to an IS evaluation at a single case study organisation. This approach is essentially a mind map representation of a system or set of causal statements, within a situational context, that could effectively be enumerated in terms of a simulation algorithm.

These studies indicate that adopting alternative non-accounting-based measures of IT-enabled benefits can be useful for researchers, especially when attempting to capture intangible benefit levels. However, Murphy and Simon (2002) caution that it is important to ensure that any proxy measures are agreed with business managers, to ensure that they are considered valid throughout the business. The work of Klein et al. (1997) suggests that as well as considering which measures are most appropriate for particular benefits, the nature of the technical functionality should also be considered. Objective, accounting measures may be preferable for systems that are concerned with transaction processing and thereby likely to deliver tangible benefits. However, systems that are designed to improve decision-making and have less tangible organisational impacts require perceptual-based measures or proxy measures to be able to effectively assess the delivery of business benefits. Consequently, Tallon and Kraemer (2007) recommend the use of both objective and perceptual data when assessing IS/IT impacts.

16.6   Benefits Measurement in Practice

Although most researchers have attempted to independently measure the benefits of organisational systems, others have attempted to assess the benefits management practices of IS professionals and a number have gathered data on current benefits measurement practices in use. For example, Berghout et al. (2011) conducted eight case studies of financial service organisations in the Netherlands to investigate the adoption of lifecycle cost and benefit management practices. They report that financial measures (e.g. return on investment) or nonfinancial indicators (e.g. achieving a strategic match) were used for IS/IT investment justification. However, they found that the cost/benefit analyses of almost all of their case study organisations were incomplete. Further, in most justification processes, investment goals were not set in a way to allow evaluation afterwards. There was a tendency to formulate qualitative ('better than the current situation') rather than quantitative goals ('an improvement of 15 %'), which would complicate evaluation. They also report that there was little proactive measurement of benefits during the realisation or exploitation stages of a project with few organisations adopting service-level agreements to evaluate performance between the IT department and users or linking operation costs and business value. They conclude that few organisations conducted an evaluation of their IS/IT investments largely because of setting nonmeasurable goals and the difficulty of isolating the influences of other investments.

Slightly better experiences are reported by Gacenga et al. (2010) in their study of IT service management benefits and performance in Australian firms. They found that the balanced scorecard was the most popular measurement technique adopted by managers supported by metrics at the process level. However, they also report that many of their respondents reported difficulties in measuring and reporting benefits in their organisations, which Gacenga et al. explain through a lack of adoption of performance measurement frameworks. A further study, also conducted in Australia, has examined the uptake of benefits management and evaluation practices (Love et al. 2005). Love et al. (2005) develop benchmarking metrics at three benefit levels (strategic, tactical and operational) for SMEs implementing IS/IT investments. They also argue for the inclusion of benchmarking metrics for risk factors associated with IS/IT implementations (such as reluctance of employees to adapt to change) that may inhibit the delivery of benefits. They conclude that strategic benefits from IS/IT investments vary across industry sector and claim to have developed measures that should allow SMEs to assess their performance against similar organisations. However, Love et al. (2005) only apply these measures to assess the overall performance of SMEs in Australia rather than for application at the project level for individual SMEs. They also do not provide guidance on how these items should be measured for individual IS/IT investments, identifying several items that were considered intangible and therefore difficult to measure (Ashurst et al. 2008). Consequently, it would appear that many IS practitioners are not applying benefits measurement in practice, possibly because the challenges of benefits measurement have yet to be resolved.

16.7 Conclusions: The Future of Benefits-Oriented Research

In his much-cited book, *The Information Paradox*, Thorp (1998) emphasises the critical importance of measuring of IT-enabled benefits for ensuring the effectiveness of the benefits management processes stating *'if you can't measure it, you can't manage it'*. Our study, which has examined and synthesised the benefits measurement literature, has raised some important concerns about the ability of organisations to measure the benefits being realised from their IT investments, and it brings into question their ability to effectively manage the realisation of benefits. More specifically, the studies of the broad benefits measurement approaches adopted by practitioners indicate that there are significant deficiencies in their current practices: too often benefits are estimated at the outset of a project but they are not systematically reappraised once the system has gone operational (Ashurst et al. 2008). The ability of organisations to effectively measure the benefits of IT is also called into question by the studies that have attempted to measure the relationship between IS/IT investment and the business value realised: the very variable results produced by these studies suggest that many organisations are neither measuring nor managing the benefits of their IT investments effectively.

In addition to highlighting serious concerns about the practice of benefits measurement, this study has also identified some significant gaps in the literature. Whilst there have been many studies that have attempted to model the IT value relationship or explored the broad evaluation approaches being adopted, there has been a dearth of research that explicitly investigates how these measures are actually operationalised within organisations and how effective they are in supporting practitioners to monitor and manage the process of IS-/IT-enabled benefits realisation. This is a very important oversight, because unless we know how benefits are currently being measured, it will be very difficult to discern how these approaches can be improved. Consequently, despite considerable interest among industry and academia in leveraging benefits and value from IS/IT investments, good results are not going to be consistently achieved without further research into the development of reliable and accepted measures. Against this backdrop, it is important that a new research agenda be established for benefits measurement, to ensure that these significant gaps are filled, as a matter of some urgency. Key issues on this agenda would include:

- Empirical studies to investigate the application of benefits measurement techniques at the process and project levels and how these methods can be used to inform adjustive action and regular monitoring of benefits realisation. Such studies would be valuable, as they will help to bridge the theory-practice gap in benefits management (Ashurst et al. 2008).
- Studies to investigate how objective and perceptual measures can be effectively combined to proactively measure and monitor the delivery of benefits during and following the completion of systems development projects (Tallon and Kraemer 2007). In particular, it would be valuable to investigate how such measures could be embedded in existing systems development methods to ensure that they become a core aspect of IS/IT projects rather than a nice idea that is put to one side once the business case is accepted.
- Research that involves longitudinal data would also be helpful to assess the strengths, limitations and accuracy of existing and new benefits measurement techniques. Research methods such as case study or action research would be particularly pertinent for these studies to get a richer interpretation of the benefits journey for both academics and practitioners and thereby increase the likelihood of realising business benefits from IS/IT investments (Doherty et al. 2012).

To conclude, if the unacceptable high levels of information systems failure and underperformance are to be successfully addressed, then it is essential that members of the academic and practitioner communities actively engage with this agenda sooner, rather than later.

References

Anthony RN (1965) Planning and control systems: a framework for analysis. Harvard University, Boston, MA

Argyropoulou M, Ioannou G, Koufopoulos DN, Motwani J (2009) Measuring the impact of an ERP project at SMEs: a framework and empirical investigation. Int J Enterprise Inform Syst 5 (3):1–13

Ashurst C, Doherty NF, Peppard J (2008) Improving the impact of IT development projects: the benefits realization capability model. Eur J Inform Syst 17(4):352–370

Barua CHK, Mukhopadhyay T (1995) Information technologies and business value - an analytic and empirical-investigation. Inform Syst Res 6(1):3–23

Benaroch M, Kauffman RJ (1999) A case for using real options pricing analysis to evaluate information technology project investments. Inform Syst Res 10(1):70–86

Berghout E, Nijland M, Powell P (2011) Management of lifecycle costs and benefits: lessons from information systems practice. Comp Indus 62(7):755–764

Bradley G (2006) Benefit realisation management. Gower, London

Bresnahan TF, Brynjolfsson E, Hitt LM (2002) Information technology, workplace organization, and the demand for skilled labor: firm-level evidence. J Econ 117(1):339–376

Brynjolfsson E, Hitt LM (1998) Beyond the productivity paradox. Commun ACM 41(8):49–55

Brynjolfsson E, Hitt LM (2000) Beyond computation: Information technology, organizational transformation and business performance. J Econ Perspect 23–48

Buccoliero L, Calciolari S, Marsilio M (2008) A methodological and operative framework for the evaluation of an e-health project. Int J Health Planning Manag 23(1):3–20

Changchit C, Joshi KD, Lederer AL (1998) Process and reality in information systems benefit analysis. Inform Syst J 8(2):145–162

Doherty NF, Ashurst C, Peppard J (2012) Factors affecting the successful realisation of benefits from systems development projects: findings from three case studies. J Inform Technol 27 (1):1–16

Farbey B, Land F, Targett D (1993) How to assess your IT investment: a study of methods and practice. Butterworth-Heinemann, Oxford

Fox S (2008) Evaluating potential investments in new technologies: Balancing assessments of potential benefits with assessments of potential disbenefits, reliability and utilization. Critic Perspect Account 19(8):1197–1218

Gacenga F, Cater-Steel A, Toleman M (2010) An International analysis of IT service management benefits and performance measurement. J Glob Informat Technol Manag 13(4):28–63

Giaglis GM, Mylonopoulos N, Doukidis GI (1999) The ISSUE methodology for quantifying benefits from information systems. Logistics Inform Manag 12(1/2):50–62

Gorla N, Somers TM, Wong B (2010) Organizational impact of system quality, information quality, and service quality. J Strat Inform Syst 19(3):207–228

Gregor S, Martin M, Fernandez W, Stern S, Vitale M (2006) The transformational dimension in the realization of business value from information technology. J Strat Inform Syst 15 (3):249–270

Hitt LM, Wu DJ, Zhou XG (2002) Investment in enterprise resource planning: business impact and productivity measures. J Manag Inform Syst 19(1):71–98

Jeffers PI, Muhanna WA, Nault BR (2008) Information technology and process performance: an empirical investigation of the interaction between IT and Non-IT resources. Dec Sci 39 (4):703–735

Klein G, Jiang JJ, Balloun J (1997) Information system evaluation by system typology. J Syst Software 37(3):181–186

Kleist VF (2003) An approach to evaluating E-business information systems projects. Inform Syst Front 5(3):249–263

Lin WT (2009) The business value of information technology as measured by technical efficiency: evidence from country-level data. Dec Support Syst 46(4):865–875

Lin WT, Chuang CH, Choi JH (2010) A partial adjustment approach to evaluating and measuring the business value of information technology. Int J Prod Econ 127(1):158–172

Love PED, Irani Z, Standing C, Lin C, Burn JM (2005) The enigma of evaluation: benefits, costs and risks of IT in Australian small-medium-sized enterprises. Inform Manag 42(7):947–964

Marchand D, Peppard J (2008) Designed to fail: why IT projects underachieve and what to do about it. IMD Int, 11 p 30

Martin-Oliver A, Salas-Fumas V (2012) IT assets, organization capital and market power: contributions to business value. Dec Support Syst 52(3):612–624

Mintzberg H (1983) Structures in fives. Designing Effective Organizations, Prentice-Hall

Mirani R, Lederer AL (1998) An instrument for assessing the organizational benefits of IS projects. Dec Sci 29(4):803–838

Mukhopadhyay T, Kekre S (2002) Strategic and operational benefits of electronic integration in B2B procurement processes. Manag Sci 48(10):1301–1314

Murphy KE, Simon SJ (2002) Intangible benefits valuation in ERP projects. Inform Syst J 12(4):301–320

Nevo S, Wade MR (2010) The formation and value of it-enabled resources: antecedents and consequences of synergistic relationships. MIS Quart 34(1):163–183

Petter S, DeLone W, McLean ER (2012) The past, present, and future of IS success. J Assoc Inform Syst 13(5):2

Powell TC, Dent-Micallef A (1997) Information technology as competitive advantage: the role of human, business, and technology resources. Strat Manag J 18(5):375–405

Ray G, Barney JB, Muhanna WA (2004) Capabilities, business processes, and competitive advantage: choosing the dependent variable in empirical tests of the resource-based view. Strat Manag J 25(1):23–37

Rehesaar H, Mead A (2005) An extension of benefit cost analysis to IS/IT investments. Bus Rev Cambridge 4(1):89–93

Remenyi D, Twite A, Money A (1993) Guide to measuring and managing IT benefits. Blackwell Publishers, New York, NY

Remenyi D, White T, Sherwood-Smith M (1997) Achieving maximum value from information systems: a process approach. Wiley, New York, NY

Sanchez H, Robert B (2010) Measuring portfolio strategic performance using key performance indicators. Project Manag J 41(5):64–73

Shang S, Seddon PB (2002) Assessing and managing the benefits of enterprise systems: the business manager's perspective. Inform Syst J 12(4):271–299

Sharif AM, Irani Z (2006) Exploring fuzzy cognitive mapping for IS evaluation. Eur J Operat Res 173(3):1175–1188

Skok W, Kophamel A, Richardson I (2001) Diagnosing information systems success: importance-performance maps in the health club industry. Inform Manag 38(7):409–419

Smith HA, McKeen JD (1993) How does information technology affect business value - a reassessment and research propositions. Revue Canadienne Des Sciences De L Administration-Can J Admin Sci 10(3):229–240

Tallon PP, Kraemer KL (2007) Fact or fiction? A sense making perspective on the reality behind executives' perceptions of IT business value. J Manag Inform Syst 24(1):13–54

Thorp J (1998) The information paradox - realising the business benefits of information technology. McGraw-Hill, Toronto

Ward JL, Daniel E (2006) Benefits management: delivering value from IS and IT investments. Wiley, New York, NY

Ward J, Elvin R (1999) A new framework for managing IT-enabled business change. Inform Syst J 9(3):197–221

Williams D, Parr T (2006) Enterprise programme management: delivering value. Macmillan, Palgrave

Yu JH, Lee HS, Kim W (2006) Evaluation model for information systems benefits in construction management processes. J Construct Eng Manag 132(10):1114–1121

Index

A

Access networks, 134
Accounting-oriented measures, 263
Accurate locational data, 239
ACO. *See* Ant colony optimisation (ACO)
Adaptation, 257
Adaptive organisation, 38
Adaptive query, 126–129
Agenda for improvements, 161
Agent-based simulation, 183
Agile development, 24
Agility, 8
Analysis of perceptions, 160
Analytics, 85–99
    tools, viii
Ant colony optimisation (ACO), 129, 135
Apache SOLR, 120
ARIMA, 73
Artificial intelligence (AI), 107, 179
Asset, 16, 199, 213–232, 251
Asset replacement, 217
Attitudes, 200, 201, 211
Automated learning, 129
Automation, 22, 23, 26, 104–106, 244
Automotive, 251
Autonomous knowledge workers, 205
Autonomy, 202, 208
Availability, 20

B

Backlog, 38
Barrier, 8, 202
Behaviour, 201, 203, 211, 221
Behavioural evidence, 201
Behavioural science, 19

Behavioural studies, 201
Benchmark, 240
Benchmarking, 265
Benefits, 80–82, 243, 257–266
    evaluation, 260
    exploitation, 9
    management, 257–266
    measurement, 257–266
    realization, 266
Big data, ix
BPEL, 87
British Telecom (BT), 113, 184, 185
Budget, 8
Business
    benefits, 3
    change, 3
    environment, 19
    improvement, 87
    performance, 206
        metrics, 34
    priorities, 26
    process, 6, 32, 35, 85, 90, 98, 159
        design, 10
        improvement, 88
        performance, 98
    transformation, viii, 3–11, 161
        project, 6
Business process model and notation (BPMN), 86
Business-wide measures, 207

C

Capabilities, 10
Capacity
    imbalances, 103
    reservations, 20